Stamp Duty Land Tax Handbook

Second edition

Chris Hart
Partner, Hart Consulting
Consultant to the Professional Tax Practice
Former Chairman, RICS Taxation Policy Panel

Tony Johnson
Before he passed away in 2007, he was a
Consultant to Edwin Hill, Chartered Surveyors
Visiting Professor, City University
Honorary Fellow, College of Estate Management
Former Chairman, RICS Taxation Standing Working Party

2009

Routledge
Taylor & Francis Group
LONDON AND NEW YORK

First published 2009 by Estates Gazette

Published 2014 by Routledge
2 Park Square, Milton Park, Abingdon, Oxfordshire OX14 4RN
711 Third Avenue, New York, NY 10017, USA

First issued in hardback 2017

Routledge is an imprint of the Taylor & Francis Group, an informa business

ISBN 13: 978-1-138-16342-3 (hbk)
ISBN 13: 978-0-7282-0525-3 (pbk)

Cover design by Rebecca Caro
Typeset in Palatino 10/11 by Amy Boyle

Contents

Preface

When stamp duty land tax was first proposed, Tony Johnson and I put together a half day conference package which we offered at several venues around the country.

Delegates raised many questions which tested our understanding of the tax to the limit, and sometimes beyond! As we researched the answers, we felt that we had a growing understanding of the subject. This led us to put it down in writing, in this book.

The late Tony Johnson and I are chartered surveyors and valuers, not tax advisors. However, this is a property tax, and our hope is that readers will benefit from an interpretation of the tax from the viewpoint of surveyors active in the property and property taxation world. For example, such concepts as net present value in respect of leases have held no fear or problems for us as valuers, notwithstanding the totally unrealistic assumptions imposed by the legislators.

It is now a year since Tony Johnson passed away after a relatively short but severe period of illness. We had discussed the second edition of this book prior to his last illness and it was Tony's opinion that few property taxes had been subject to such frequent alteration and amendment. As Richard Hayward wrote shortly after Tony's death:

> he was the author of many professional articles and most notably *Modern Methods of Valuation*; generally regarded as the standard valuation texts for students. At the time of his death he was still writing; this time collaborating in a book on Stamp Duty Land Tax. His other published works included books on Betterment Levy, Development Land Tax and other taxation legislation, all of which were subsequently repealed. Indeed, as he used to observe on occasions, perhaps he should have set out to write a book on Income Tax! He also edited the Estates Gazette Diary for many years.

From my own perspective I now know how much easier it is to produce a book with Tony's help and guidance than it is without, this second edition is dedicated to his memory.

Chris Hart
Stratford upon Avon
August 2008

Table of Cases

Table of Statutes

Definitions

i. Major interest in land — (s 117 of Finance Act 2003 (FA 2003))

For the purposes of stamp duty land tax (SDLT), in England and Wales it is an estate in fee simple absolute, or a term of years absolute, which subsist in law or equity.

In Scotland, it is the interest of an owner of land or the tenant's right over or interest in a property subject to a lease. (At present an interest in relation to a feudal estate is that of the proprietor of the dominium utile.)

In Northern Ireland, it is any freehold estate or any leasehold estate.

ii. Land — (s 121 of FA 2003)

Land includes buildings and structures, and land covered by water.

iii. Residential property — (s 116 of FA 2003)

This is defined as:

a. a building that is used or suitable for use as a dwelling, or is in the process of being constructed or adapted for such use
b. land that is or forms part of the garden or grounds of a building within (a) (including any building or structure on such land)

c. an interest in or right over land that subsists for the benefit of a building within paragraph (a) or of land within paragraph (b).

Non-residential property is any property that is not residential property, or six or more residential properties in a single transaction. The meaning of this definition is considered in detail in Chapter 2.

iv. Market value — (s 118 of FA 2003)

Market value "in relation to any assets means the price which those assets might reasonably be expected to fetch on a sale in the open market". The definition comes direct from the Taxation of Chargeable Gains Act 1992, and brings with it its own complex form of application; this is dealt with in detail in Chapter 10.

v. The effective date of a land transaction — (s 119 of FA 2003)

Except as otherwise provided, the effective date is the date of completion. The main exceptions are where a contract is substantially performed (see Chapter 1) before completion or without being completed (ss 44 and 119 of FA 2003). For options and rights of pre-emption it is the acquisition date, not the date they become exercisable.

A contract on its own is not a land transaction until either completion of the transaction, or its substantial performance, creates the effective date. Where there are unascertainable future events at the initial effective date that may lead to a further payment of consideration or of rent and which are unquantifiable at that time, then it is the point at which the additional payment becomes quantifiable that creates a further effective date.

vi. Reverse premiums — (Schedule 17A to FA 2003, para 18)

For a grant, assignment, or surrender of a lease, a reverse premium does not count as chargeable consideration. The legislation is specific as to what constitutes a reverse premium and only the following qualify: in relation to the grant of a lease, a premium paid by the landlord to the tenant; in relation to an assignment, a premium paid by the assignor to the assignee; in relation to the surrender of a lease, a premium paid by the tenant to the landlord.

vii. General definitions

See s 121 for "minor definitions" and s 122 for an "index of defined expressions".

1 Liability to Stamp Duty Land Tax

A. Introduction

1.1 Stamp duty land tax (SDLT) was introduced by the Finance Act 2003 (FA 2003) and took effect from 1 December 2003 in respect of land transactions, replacing the long established stamp duty. Significant changes to the tax scheme have been made by every Finance Act since, and are incorporated in this book, as are regulatory changes, and further changes made by the Finance Act 2008.

1.2 Unlike stamp duty, which was a tax on documents, SDLT is a tax on transactions. So, if a property is sold or let without any documents recording the transaction, SDLT will, none the less, be payable.

B. Events leading to a charge

i. Land transactions

1.3 SDLT is charged on land transactions. A land transaction is "any acquisition of a chargeable interest" (s 43 of FA 2003).

ii. Chargeable interest

1.4 A chargeable interest means "an estate, interest, right or power in or over land in the United Kingdom" (s 48(1)(a) of FA 2003). It therefore includes freehold and leasehold interests (and their equivalent in Scotland). It also includes the benefit of an obligation, restriction or condition affecting the value of any such estate, interest, right or power (s 48(1) of FA 2003).

1.5 Her Majesty's Revenue and Customs' (HMRC), *Stamp Duty Land Tax Manual* states; chargeable interests in England and Wales and Northern Ireland include:

- A freehold estate (sometimes referred to as an estate in fee simple).
- A leasehold estate (sometimes referred to as a term of years).
- An undivided share in land.
- A right in or over land, such as an easement or profit a prendre.
- A rent charge.
- In Northern Ireland, a ground rent or fee farm rent.
- The right to receive rent (*IRC* v *John Lewis Properties Ltd* [2001] STC 1118).
- The benefit of a restrictive covenant.
- The benefit of a positive covenant.
- An equitable interest in land, such as a life interest, an interest in reversion or remainder.
- An executor or trustee's power of appointment (the only likely example in practice of a power over land).

Chargeable interests in Scotland include:

- Ownership of land.
- Any other heritable right in or over land.
- The tenants interest under a lease of land.
- A servitude.
- A life rent.

1.6 However, some interests, set out below, are exempt from this definition.

a. Security interest
These are interests held to secure the payment of money or the performance of an obligation. A typical example is a mortgage.

b. Licences
These are licences to use or occupy land.

c. Tenancies at will
These exist in the United Kingdom apart from Scotland, as do the following examples. As they are terminable without notice, they are little different from licences.

d. Advowson
This is the right of presentation to a beneficiary.

e. Franchises
These are grants by the Crown, such as the right to hold a market or fair, or to take tolls.

f. Manors
There are three elements to a manor:

1. Lordship of the manor. The owner may style himself lord of the manor (of, say, Keswick).
2. Manorial land. As a manor was a defined area it may include land.
3. Manorial rights. Sometimes rights are retained by the lord on disposal of part of the manor, for example, hunting, fishing or shooting rights.

Prior to the Land Registration Act 2002 (LRA 2002), these rights could be registered. Since that date there can be no application for a first registration. However, dealings with an already registered title are subject to compulsory registration, eg the granting of a lease. Under LRA 2002, manorial rights are categorised as overriding interests. These rights will lose their overriding status after 12 October 2013 under s 117 of LRA 2002.

1.7 A chargeable interest also includes "the benefit of an obligation, restriction or condition affecting the value of any such estate, interest, right or power" (s 48(1)(b) of FA 2003). Typical examples are easements and restrictive covenants. Hence the surrender of the benefits for consideration would be a land transaction attracting SDLT.

g. Partnerships

1.8 There are now a number of events where transactions involving partnerships are not treated as land transactions. Due to the comprehensive changes to the partnership regime these are dealt with in Chapter 9.

iii. Transaction by contract and conveyance

1.9 Most land transactions involve the parties entering into a contract which is completed by the actual conveyance of the interest to the purchaser or lessee.

1.10 If a payment is made at the contract stage which is not a substantial amount, this does not attract SDLT. The tax will be payable on completion, when the balance of the purchase price is paid, the contract and the conveyance being treated as parts of a single land transaction (s 44 of FA 2003). Common practice is for a payment of 10% of the purchase price to be made at contract stage, while a substantial amount would normally be taken to be 90% or more.

1.11 If the contract is substantially performed before completion, or, indeed, if there is no subsequent conveyance, the transaction will be treated as taking place when the contract is substantially performed. The purpose of this is to render ineffective the practice under the stamp duty regime of resting on a contract, whereby the parties relied on the contract alone without any conveyance. In this way no stamp duty was payable. The new approach, whereby substantially performed contracts attract the tax, illustrates the transaction based approach of SDLT, in contrast to the document based approach adopted for stamp duty.

1.12 A contract can be substantially performed in two ways.

1.13 One is where the purchaser, or a person connected with the purchaser, takes possession of the whole, or most of, the subject matter of the contract, and receives, or becomes entitled to receive rents or profits.

1.14 If the purchaser is allowed to take possession after the contract and before completion, commonly to carry out work to a property having been given the keys, this would appear to be caught by this provision since the purchaser in possession is entitled to receive rents or profits.

1.15 This provision applies equally where the vendor grants a licence or a short term lease to the purchaser allowing possession.

1.16 The second way in which a contract can be substantially performed is if a substantial amount of the consideration is paid or provided.

1.17 If none of the consideration is rent, the payment of the whole of the consideration or most of it will amount to a substantial amount. This normally applies to payments of 90% or more of the whole amount (see Land Transaction Return Guidance Notes (SDLT 6) published by HMRC, and Stamp Duty Land Tax Manual 07950).

1.18 If the whole of the consideration is rent, the contract is substantially performed when the first payment of rent is made.

1.19 If the consideration is both rent and other consideration, for example, payment of a premium under a lease plus rent, the contract is substantially performed on the earlier of payment of rent or of 90% or more of the other consideration.

iv. Subsales

1.20 The provisions regarding transactions by contract and conveyance apply where the parties to the contract or agreement are the same parties to the completion by conveyance. However, in some instances the purchaser, or lessee, will assign or sub-sell the benefits of the contract to a third party, the transferee, who is then entitled to be the party to the completion by conveyance with the vendor, or lessor (s 44A of FA 2003). In such instances the third party becomes liable to pay any SDLT that is due (s 45 of FA 2003).

1.21 It is assumed that there is a new contract, a secondary contract, under which the transferee is the purchaser, and the consideration is that payable under the original contract, together with the consideration payable to the original purchaser, or lessee, for the assignment or subsale.

Example 1.1 Assignment of contract

A agrees to sell the freehold interest in a property to B for £200,000, and contracts are exchanged. Prior to completion, B assigns his interest in the contract to C for £100,000, and requires A to convey the interest to C.

Under the contract, C will pay £200,000 to A, but, for SDLT purposes, is treated as having a contract to pay the sum under the original contract of £200,000, plus the sum payable to B of £100,000, a total of £300,000, on which sum SDLT will be assessed.

This will still be so if the payment to A is made by a person connected to C (see Appendix 3).

Example 1.2 Subsale

As for Example 1.1, but B contracts with C to sell on the interest for £300,000.

Again, it is assumed for SDLT that there is one contract under which C is the purchaser and SDLT will be assessed on the payment by C of £300,000.

C. Liability for tax

1.22 SDLT is payable by the purchaser, which includes the lessee in the case of a lease. Provisions for the liability to pay the tax are contained in s 85 and s 86 of FA 2003. There are special provisions in certain cases.

i. Joint purchasers

1.23 Where there are two or more purchasers of an interest, the obligation to pay SDLT is a joint obligation but which may be discharged by any one of them. If they fail to meet the obligation, they are jointly and severally liable for the consequences (s 103 of FA 2003).

ii. Partnerships

1.24 Where a land transaction is entered into by a partnership, it is treated as entered into on behalf of the partners and not the partnership as such (s 104 of and Schedule 15 to FA 2003). A partnership includes not only an ordinary partnership within the Partnership Act 1890 but also a limited partnership and a limited liability partnership.

1.25 Hence, the obligation to pay SDLT is a joint obligation of the responsible partners, who are jointly and severally liable. Responsible partners are those who were partners at the effective date of the land transaction, or who became a partner after that date (Schedule 15 to FA 2003, para 6).

1.26 In the alternative, a majority of the partners may nominate one or more partners to be a representative partner, who will act on behalf of the responsible partners after notice of the nomination has been given to HMRC (Schedule 15, para 8) (and see Chapter 9).

iii. Trustees of settlement

1.27 Provisions regarding the role and duties of trustees are contained in s 105 of and Schedule 16 to FA 2003. In essence, trustees are treated the same as other owners so, for example, an acquisition by a trustee of a settlement (a trust which is not a bare trust) is treated as the acquisition of the whole of the interest, including the beneficial interest, and the trustee is responsible for making the land transaction return and for payment of the tax or interest on unpaid tax. (An acquisition by a bare trustee is treated as if it were the act of the person for whom they are the trustee.)

iv. Unit trust schemes

1.28 In the case of a unit trust scheme, the trustees are treated as if they were a company (see para 1.32) and the rights of the unit holders as if they were shares in the company. A unit trust scheme has the same meaning as in the Financial Services and Markets Act 2000 (s 101 of FA 2003).

1.29 A unit trust scheme is not treated as a company for the purposes of s 53 of FA 2003 (deemed market value rule for connected companies — see para 4.3 and Chapter 16) and of Schedule 7 (group, reconstruction and acquisition relief — see para 1.57 and Appendix 4).

1.30 In the case of an umbrella scheme, each part of the scheme is treated as a separate unit trust, rather than the scheme as a whole. An umbrella scheme is one which provides for separate pooling of participants' contributions and the profits or income out of which payments are to be made by them, and where the participants are entitled to exchange rights in one pool for those in another.

v. Persons acting as representatives for others

1.31 A person who has the direction, management or control of an incapacitated person is responsible for the payment of SDLT on a purchase on behalf of that incapacitated person, but can recover the sum paid from money received on behalf of the incapacitated person (s 106(1) of FA 2003).

1.32 Similarly, the personal representatives of a deceased person are liable for SDLT on a purchase concluded after the death, but can recover the payment out of the deceased's estate (s 106(3)).

1.33 The parent or guardian of a minor is responsible for the payment of SDLT, on a purchase on behalf of the minor, which has not been paid by the minor (s 106(2)), as is a receiver appointed by a court who has the direction and control of a property (s 106(4)).

1.34 From 2008, a person acting as agent for a taxpayer may complete, and sign, a form of self assessment (SDLT 60) see also Chapter 11.

vi. Companies

1.35 Companies are defined in s 100 of FA 2003 as any body corporate or unincorporated association, but do not include a partnership (for which see Chapter 9). A company must act through its "proper officer" or someone with the express or implied authority to act on its behalf. In the case of a body corporate, the proper officer is the company secretary or acting company secretary. If there is no such post, the proper officer is the company treasurer or acting company treasurer, who is also the proper officer of an unincorporated association. The situation differs where a liquidator or administrator is involved, where they effectively become the proper officer.

1.36 Open-ended investment companies are defined in s 102 of FA 2003 and have the meaning ascribed by s 236 of the Financial Services and Markets Act 2000. Regulations may be made by the Treasury to ensure that the effect of the legislation, in relation to open-ended companies, matches that for unit trust schemes, so that company assets in separate pools have the same treatment as separate pooling under an umbrella scheme (see para 1.28).

D. Exempt transactions

i. General

1.37 There are a considerable number of transactions which do not attract a charge to SDLT, some are exemptions while others are reliefs. They are contained in ss 57A to 75 of FA 2003 and Schedules 3 and 6A. They comprise the following categories.

ii. No consideration

1.38 If there is no chargeable consideration then no SDLT is payable, for example a land transaction which is an outright gift (s 49 of and Schedule 3 to FA 2003, para 1, and see Chapter 4 for what constitutes chargeable consideration). However, some gifts to companies may attract a charge to SDLT, see Chapter 16 and Appendix 4.

iii. Residential property exchanges with house building companies

1.39 Where a house building company purchases a dwelling owned by an individual who is purchasing one of the company's new dwellings, the purchase by the company is treated as being for no consideration (see Schedule 6A to FA 2003, para 1 (inserted by s 58A)).

1.40 Certain conditions apply for the exemption to operate. These are:

a. The purchase must be by a house building company or by a company connected with it. A house building company is one that carries on the business of constructing or adapting buildings, or parts of buildings, for use as a dwelling.

b. The individual, either alone or with others, is purchasing a new dwelling from the company. A new dwelling can be a dwelling constructed for use as a single dwelling, or a dwelling adapted for use as a single dwelling, and which has not been previously occupied since it was constructed or adapted.

c. The dwelling sold by the individual was occupied as the individual's only or main residence for some time in the two years prior to the sale, and the intention is for the new dwelling to be occupied as the only or main residence in its place. The identification of a dwelling as being a person's only or main residence is necessary for the relief from capital gains tax (CGT) given by s 222 of the Taxation of Chargeable Gains Act 1992. It is assumed that the test of whether a dwelling is such a residence will follow the CGT approach. It cannot be identical since, for example, CGT relief does not require actual residence within three years prior to the transfer. Each transaction is in consideration of the other.

d. The area of land being purchased with the dwelling by the company does not exceed the permitted area. This is a concept also derived from CGT legislation and is considered in more detail in Chapter 2. It is, broadly, garden and grounds with an area not exceeding 0.5 ha or, if more, the land required for the reasonable enjoyment of the dwelling as a dwelling having regard to its size and character. If the land does exceed the permitted area, the company will be liable for SDLT on the difference between the price paid and the market value of the dwelling if standing in grounds not exceeding the permitted area. In most cases where this applies, it is probable that the difference between the price and the market value will be within the exempt limit (see Chapter 4).

iv. Residential property exchanges with property traders

1.41 Similar provisions to those for residential property exchanges with house builders apply where the dwelling owned by an individual is purchased by a property trader that carries on the business of buying and selling dwellings (Schedule 6A, para 2 of the FA 2003). A property trader is a company or a limited liability partnership (LLP) or a general partnership of companies and/or LLPs (Schedule 6A, para 8).

1.42 The requirements in respect of the individual selling the dwelling are the same as for the exchange of property with a house building company. Hence, the individual will purchase a new dwelling from a house building company, but not from the property trader.

1.43 As to the conditions for the property trader, these are that there is no intention to spend more than the permitted amount on refurbishment on the dwelling purchased from the individual, nor to grant a lease or licence of the dwelling (apart from up to six months to the vendor), nor to permit any of the principals or employees or persons connected with them to occupy the dwelling. A further condition is that the property trader is making the acquisition in the course of a business that consists of or includes acquiring dwellings from individuals who themselves acquire new dwellings from house building companies.

1.44 Refurbishment of a dwelling means works which are intended to enhance the value of the dwelling, but do not include cleaning the dwelling or works required to ensure that it meets minimum safety standards. The permitted amount which may be spent on the dwelling is the greater of £10,000 or 5% of the consideration for the acquisition, but subject to a maximum amount of £20,000.

1.45 Similar provisions apply where a property trader acquires a dwelling from the personal representatives of a deceased individual. The conditions are that the acquisition is made in the course of a business that consists of or includes acquiring dwellings from personal representatives of deceased individuals, and that the deceased individual occupied the dwelling as their only or main residence at some time in the period of two years ending with the date of their death. The provisions regarding the permitted amounts and grant of lease or licences also apply as before (Schedule 6A, para 3).

1.46 A third situation where a property trader will not be liable for payment of SDLT is where it acts as a chain breaker. This will arise where an individual has arranged to sell a dwelling and to acquire a new dwelling and the arrangements for the sale of the existing dwelling fail, so making the individual unable to proceed with the purchase of the new dwelling. The conditions are that the trader acquires the interest in the old dwelling and the acquisition is made in the course of a business that consists of or includes acquiring dwellings from individuals in those circumstances, and the individual occupied the dwelling as their main residence at some time in the last two years before the acquisition and intends to occupy the new dwelling as their main residence. The provisions regarding permitted amounts and the granting of a lease or licence, etc also apply as before (Schedule 6A, para 4).

v. Purchase of employee's dwelling

1.47 Where an employer purchases a dwelling from an employee, the acquisition will be exempt from tax, similar to the exemption given to a house building company (Schedule 6A, para 5 of the FA 2003).

1.48 The conditions for the exemption to apply are:

a. The employee occupied the dwelling as his or her sole or main residence for the whole or part of the two years prior to the date of purchase. The meaning of main residence is considered in para 1.40 above.

b. Relocation of employment may be due to the individual becoming an employee of that employer, or the employee's duties changing, or the employee working in a different work place. A claim that the change of residence results from the relocation of employment must show that the change "is made wholly or mainly to allow the individual to have his residence within a reasonable daily travelling distance of his new place of employment, and his former residence is not within a reasonable daily travelling distance of that place" (Schedule 6A to FA 2003, para 5(5)).

c. The consideration paid by the employer does not exceed the market value of the dwelling. Where the area of land with the dwelling exceeds the permitted area, the chargeable consideration is determined as for a purchase by a house building company (see 1.40 above). If further sums are paid such as removal costs or financial inducements, HMRC may question whether these are part of the purchase price, taking it above market value.

d. The area of land does not exceed the permitted area (Schedule 6A, para 7 and see para 1.40 for meaning of permitted area).

1.49 This exemption given to employers is also given to a property trader in these circumstances where it acts on behalf of the employer in the change of residence of the employee (Schedule 6A, para 6). A relocation company is one which provides the service of acquiring dwellings in these circumstances, or a company which is connected to such a company.

1.50 In the case of the exemption given to property traders outlined above, the relief from payment of SDLT will be withdrawn in each case if the property trader spends more than the permitted amount on refurbishment of the old dwelling, or grants a lease or licence of the old dwelling, or permits any of its principals or employees or persons connected with them to occupy the old dwelling.

vi. Compulsory purchase facilitating development

1.51 A purchase under compulsory purchase powers is, normally, treated for SDLT purposes as a purchase like any other. However, there is an exception where the body exercising such powers is using those powers to facilitate development by another.

1.52 Commonly, a compulsory purchase order (CPO) is made under town planning legislation, in England and Wales under s 226 of the Town and Country Planning Act 1990, and under equivalent legislation in Scotland and in Northern Ireland.

1.53 The exercise of these powers is, generally, to enable large scale schemes to go ahead, where a large number of interests need to be purchased. In the absence of a CPO, the development could be thwarted by a few landowners. Such schemes are, typically, regeneration schemes and town centre redevelopment.

1.54 In effect, the authority exercising the CPO powers is an agent for the developer, since the authority will purchase an interest and then pass it on to the developer. Indeed, in practice, owners of affected interests will often deal directly with the developer, in the knowledge that they have to sell.

1.55 Consequently, a purchase of an interest by an authority having CPO powers after it has made a CPO or, in Northern Ireland, a purchase by means of a vesting order, where the purchase is to facilitate development by another, is exempt from SDLT (s 60 of FA 2003). The making of a CPO is, of course, the preliminary step in the CPO process and before CPO powers actually exist. Hence, a purchase by agreement in advance of the confirmation of the order is exempt.

vii. Compliance with a planning obligation

1.56 A land transaction entered into in order to comply with a planning obligation or the modification of a planning obligation is exempt from SDLT where:

a. the planning obligation is enforceable against the vendor
b. the purchaser is a public authority
c. the transaction takes place within five years from the date the planning obligation was entered into (s 61 of FA 2003).

1.57 A planning obligation is:

a. in England and Wales, an obligation entered into in accordance with s 106(9) or s 299A(2) of the FA 2003, or a modification within s 106A(1) of the Town and Country Planning Act 1990

b. in Scotland, an agreement made under s 75 or s 246 of the Town and Country Planning (Scotland) Act 1997

c. in Northern Ireland, an agreement within article 40 or a modification as mentioned in article 40A(1) of the Planning (Northern Ireland) Order 1991/1220 (NI 11).

1.58 A public authority includes central government departments, local authorities, health authorities and other planning authorities, as listed in s 61(3) of FA 2003.

1.59 An example of such a transaction would be in respect of an obligation by a developer to make land available by sale to a public authority for a public purpose, such as a road, or open space, or a school.

viii. Group relief and reconstruction or acquisition relief

1.60 Part 1 of Schedule 7 to FA 2003 (inserted by s 62) provides relief for certain transactions between group companies. In general, a transaction where the vendor and the purchaser are companies which are members of the same group is exempt from SDLT (but see Chapter 18 and Appendix 4). The test of whether companies are in the same group is, broadly, based on the 75% of ownership test (see Schedule 7 to FA 2003, para 1). This area of SDLT has been subject to significant amendment over time.

1.61 Schedule 7 further provides for exclusion of the exemption in certain circumstances where the vendor and purchaser cease to be members of the same group following the transaction. The schedule is considered in further detail in Appendix 4.

1.62 Part 2 of Schedule 7 provides relief for transactions where the purchaser company acquires the vendor company in pursuance of a scheme for the reconstruction of the vendor company. Any land transaction as part of the scheme is exempt from SDLT (reconstruction relief) if:

a. The consideration consists wholly of the issue of non-redeemable shares, or partly of such shares plus the assumption or discharge of liabilities of the vendor company.

b. Each shareholder of one company is a shareholder of the other company, holding broadly the same proportion of the total shares in each company.

c. The scheme is for commercial reasons and not for the main purpose, or for one of the main purposes, of tax avoidance.

1.63 There are further provisions for SDLT to be charged at 0.5% (the stamp duty rate) on land transactions (acquisition relief) which are part of the acquisition by a company of the whole or part of the undertaking of another company, if:

a. The consideration consists wholly of the issue of non-redeemable shares, or partly of such shares if the balance is cash not exceeding 10% of the nominal value of the shares, or is the assumption or discharge of the liabilities of the vendor, or both.

b. The purchaser company is not associated with another company which is party to arrangements with the vendor company relating to shares of the purchaser company, issued in connection with the transfer of the undertaking. Companies are associated if one controls the other, or they are both controlled by the same person.

1.64 Part 2 further provides for withdrawal of reconstruction relief and acquisition relief in certain circumstances where changes occur in the ownership of the purchaser company (see Appendix 4 for further details).

ix. Demutualisation of insurance company or building society

1.65 A land transaction is exempt from SDLT if it is for the purpose of the transfer of the whole or part of the business of a mutual insurance company to a company that has share capital, subject to conditions contained in s 63 of FA 2003.

1.66 Similarly, a land transaction is exempt if it is part of the transfer of a building society's business to a commercial company (s 64 of FA 2003).

x. Transfer of assets to trustees of unit trust scheme

1.67 The acquisition of a chargeable interest by trustees of a unit trust scheme is exempt from SDLT if:

a. Immediately before the acquisition, no assets were held by the trustees for the purposes of the unit trust scheme.

b. No units of the scheme were in issue before the acquisition.

c. The only consideration for the acquisition is the issue of units in the scheme to the vendor.

d. Immediately after the acquisition the vendor is the only unit holder of the scheme (s 64A of FA 2003).

xi. Incorporation of limited liability partnership (LLP)

1.68 The transfer of a chargeable interest to a LLP in connection with its incorporation is exempt from SDLT if:

a. The transaction takes place not more than one year after the incorporation of the LLP.

b. The transferor, at the relevant time, is a partner in the partnership, all of whose partners, but no one else, are or are to be members of the LLP, or who holds the chargeable interest as nominee or bare trustee for one or more of the other partners. The relevant time is immediately before incorporation of the LLP or, where the interest was acquired after incorporation, immediately after it was acquired.

c. The partner's share in the chargeable interest is the same before and after the transfer or, if different, not as the result of a tax avoidance scheme (s 65 of FA 2003).

1.69 A LLP is one formed under the Limited Liability Partnerships Act 2000 or the Limited Liability Partnerships Act (Northern Ireland) 2002.

xii. Transfers involving public bodies

1.70 A land transaction where the vendor and purchaser are both public bodies is exempt if it relates to a statutory reorganisation (s 66 of FA 2003). Such a reorganisation means changes involving the establishment, reform or abolition of public bodies, or the transfer of functions from one body to another.

1.71 The Treasury may, by order, provide for a land transaction effected under a statutory provision to be exempt where one of the parties is a public body. Public bodies include government offices and parliamentary institutions, local government bodies, health authorities, planning authorities, and statutory bodies. A full list is set out in s 66(4) of FA 2003.

1.72 From the FA 2007, transfers of surplus school land between public bodies were brought within the scope of the relief at s 66 of FA 2003. Previously, the relief was available under other legislation.

xiii. Parliamentary constituencies reorganisation

1.73 Where a local constituency association transfers a chargeable interest to a new constituency association as a result of changes to the area of a parliamentary constituency, such a transfer is exempt (s 67 of FA 2003).

1.74 A local constituency association is an unincorporated association formed to further the aims of a political party in that constituency.

xiv. Right to buy, shared ownership leases, shared ownership trusts, and rent to mortgage

a. Right to buy

1.75 A right to buy transaction is one where the purchaser or lessee of a dwelling acquires the interest in the dwelling at a discount to market value. The acquisition is from a public sector body, or from a private sector vendor who purchased the interest from a public sector body, where the right of the occupier to buy was preserved by s 171A of the Housing Act 1985 or s 61 of the Housing (Scotland) Act 1987 (see also Chapter 17).

1.76 A right to buy transaction provides for a further payment to the vendor if the purchaser or lessee sells or assigns within a certain period. This clawback provision represents a contingent consideration, but which is left out of account for calculation of the chargeable consideration (Schedule 9 to FA 2003, inserted by s 70, and see para 4.19 below).

1.77 A public sector body includes government departments, local authorities, social housing providers, police authorities, and others set out in Schedule 9 to FA 2003, para 1(3) (see also Chapter 17). The relief was extended in the FA 2007 to include shared ownership trusts which themselves were introduced to avoid problems with commonholds. It applies to the same categories of qualifying bodies as do the provisions concerning shared ownership leases.

b. Shared ownership lease

1.78 Certain bodies are empowered to grant leases under which the lessee has the right to acquire the reversion, a shared ownership lease. Concessions on the payment of stamp duty or the purchase of the reversion, contained in s 97 of the Finance Act 1980, are continued for SDLT by Schedule 9 to FA 2003 (see also Chapter 17).

1.79 Certain conditions, as set out in para 2(2) of the Schedule, must be met. The lease must:

a. be of a dwelling
b. give the lessee exclusive use of the dwelling
c. provide for the lessee to acquire the reversion
d. be granted in consideration partly of rent and partly of a premium calculated by reference to either the market value of the dwelling or a sum calculated by reference to that value
e. contain a statement of the market value of the dwelling or the sum calculated by reference to that value by reference to which the premium is calculated.

1.80 The lessee may elect in the land transaction return (see Chapter 11) for the chargeable consideration on the grant of the lease to be taken as the market value of the dwelling, or the sum calculated by reference to that value.

1.81 If the lessee makes such an election, the subsequent acquisition of the reversion will be exempt from SDLT.

1.82 The bodies which grant shared ownership leases to which this applies are set out in para 5 of Schedule 9 to FA 2003 and are:

a. a local housing authority
b. a housing association
c. a development corporation
d. a housing action trust
e. the Commission for the New Towns
f. the Northern Ireland Housing Executive.

1.83 Similar provisions apply to shared ownership leases where staircasing is allowed. These are leases where the lessee is entitled to make payment of a sum which will result in a reduction of the rent payable. If the conditions set out in para 1.71 are met, (save that staircasing provisions are substituted for the right to acquire the reversion, and the premium obtainable on the open market at the minimum rent payable replaces the reference to market value), no SDLT will be payable on further payments. The minimum rent is that payable following the payment of the sum.

1.84 On an election by the lessee in the land transaction return, the chargeable consideration is taken as the premium payable, or the sum calculated by reference to that sum, and also taken to be the minimum rent (see also Chapter 17).

xv. Acquisition by registered social landlord

1.85 A purchase by a registered social landlord (RSL) is exempt from SDLT if the landlord is controlled by its tenants, and either the vendor is a qualifying body, or the transaction is funded with the assistance of a public subsidy (s 71 of FA 2003).

1.86 The landlord is controlled by its tenants if the majority of the board members are tenants occupying properties owned or managed by it.

1.87 A qualifying body is one as set out in s 71(3) of FA 2003, and includes a RSL, a housing action trust, various local authorities and other public housing authorities (see also Chapter 17).

xvi. Property finance transactions

1.88 Certain transactions are entered into between individuals and financial institutions, such as sales or assignments by individuals to the institutions who lease back or sub-lease back, with the right of the individual to buy back the interest (commonly Islamic mortgages in the ijara form); or, purchases from the individual by the institution, which sells back the interest to the individual, subject to a mortgage taken by the individual from the institution (the marabaha form).

1.89 These transactions are, in essence, fund raising deals which will normally be exempt from SDLT (ss 72 and 73 of FA 2003). Exceptions include transactions where the individual is a trustee and any beneficiary is not an individual, or where the individual is a partner and any of the other partners is not an individual (see also Chapter 17 and Chapter 18).

xvii. Certain leases by RSLs

1.90 The grant of a lease of a dwelling by a RSL for an indefinite term, or terminable by notice of a month or less, is an exempt transaction if it is the result of an arrangement between the RSL and a housing authority, whereby the RSL provides temporary rental accommodation to individuals nominated by the authority in pursuance of its statutory housing functions (s 49 of and Schedule 3, para 2 to FA 2003). The RSL must itself have taken a lease of the dwelling for a period of five years or less (see also Chapter 17).

1.91 It is extremely unlikely, in any event, that any such letting by an RSL would attract a charge to SDLT on the individual.

xviii. Transactions in connection with divorce

1.92 Transactions between parties to a marriage are exempt where they are effected in pursuance of court orders made on granting a decree of divorce, nullity of a marriage and similar circumstances, and also in pursuance of an agreement of the parties in contemplation of or in connection with such matters (Schedule 3 to FA 2003, para 3). There are no similar reliefs for transactions between married persons where chargeable consideration is involved.

xix. Variation of wills

1.93 A transaction within two years after a person's death that varies a disposition under the will, or under intestacy law, without consideration other than the making of a variation of another such disposition in return, is exempt from SDLT (Schedule 3 to FA 2003, para 4).

xx. Zero-carbon properties

1.94 From the FA 2007, a new category of relief was introduced for zero-carbon homes, this relief was extended to zero-carbon flats by the FA 2008. The qualifying criteria for a property to be considered zero-carbon is set out in the Stamp Duty Land Tax (Zero-Carbon Homes) Regulations 2007 (SI 2007/3437). This adjudges the level of carbon emmissions from a property which must be zero over the course of a year. The relief is time limited until 30 September 2012, relief is given on the first £500,000 of the purchase price. If the property transferred has a value under £500,000, then the SDLT payable is nil, if the property is transferred at a higher value than £500,000, then a reduction in the applicable amount of SDLT is given in the sum of £15,000.

E. Exempt bodies

i. Charities

1.95 A purchase by a charity is exempt from SDLT (charities relief) (Schedule 8 to FA 2003, para 1, inserted by s 68). The purchase must be for qualifying charitable purposes. These are that the intention is to use the property for the furtherance of the charitable purposes of that, or another charity, or, if the property is an investment, the profits are applied to the charity's charitable purposes. The transaction must not have been entered into for the avoidance of tax by the charity or another person. There is no partial relief available, all the acquirers of the property must be charitable bodies or trusts at the point of acquisition.

1.96 There are clawback provisions if the charity ceases to be established for charitable purposes only or the property is not being used for qualifying charitable purposes (a disqualifying event) within three years after the time of the transaction, or in consequence of arrangements to cease, made before the end of the three-year period.

1.97 In such circumstances, the body must pay SDLT, which is either the sum that would have been payable in the absence of charitable relief, or an appropriate proportion thereof.

1.98 An appropriate proportion is determined by having regard to what was acquired originally and what is still owned when the disqualifying event occurs, and the extent of the property which ceases to be used for charitable purposes at the later time (Schedule 8 to FA 2003, para 2). The relief is only withdrawn if the charity or charitable trust still owns the land at the time of the disqualifying event, a disposal of the land by the charitable body does not trigger the clawback.

ii. Health service bodies

1.99 Certain health service bodies are exempt from payment of SDLT on land transactions under The Stamp Duty Land Tax (Consequential Amendment of Enactments) Regulations 2005 (SI 2005/82). They are:

a. a National Health Service trust
b. a Health and Social Services trust
c. a Primary Care Trust
d. a Local Health Board.

iii. Bodies established for national purposes

1.100 A purchase is exempt from SDLT if the purchaser is one of the following:

a. the Historic Buildings and Monuments Commission for England
b. the National Endowment for Science, Technology and the Arts
c. the Trustees of the British Museum
d. the Trustees of the National Heritage Memorial Fund

e. the Trustees of the Natural History Museum (s 69 of FA 2003).

iv. Crown

1.101 The provisions regarding SDLT apply in relation to public offices and departments of the Crown, but no payment is required if the tax would ultimately be borne by the Crown (s 107 of FA 2003). In effect, therefore, they are treated as exempt.

1.102 As to Government and Parliamentary bodies, a land transaction where any of the following is the purchaser is exempt from SDLT. They are:

Government

1. A Minister of the Crown.
2. The Scottish Ministers.
3. A Northern Ireland department.

Parliament, etc

1. The Corporate Officer of the House of Lords.
2. The Corporate Officer of the House of Commons.
3. The Scottish Parliamentary Corporate Body.
4. The Northern Ireland Assembly Commission.
5. The National Assembly for Wales.

2 Residential Property and Non-Residential Property

A. Residential buildings

i. General

2.1 In the operation of stamp duty land tax (SDLT) there is an important difference in the application of the tax between residential property and that which is not. This is particularly so in respect of designated disadvantaged areas (DDAs) which are described in Chapter 3. It may also be relevant where the consideration on purchase is below £150,000, as demonstrated in Chapter 4. Significant changes were introduced by the Finance Act 2005 (FA 2005).

2.2 The statutory definition of residential property first appeared in the FA 2002 which inserted new ss 92A and 92B into the FA 2001 after s 92, the three sections together providing for exemptions from stamp duty in DDAs, exemptions which continue for SDLT, the definition is now contained in s 116 of the FA 2003.

ii. Meaning of residential building

a. Building used as a dwelling

2.3 Where a building is actually occupied as a dwelling then it is a residential property for SDLT purposes. This would appear to be so even if such a use is contrary to planning law (s 116(1) refers to "a building that is used ... as a dwelling" without qualification as to the lawfulness of the use). It is also a residential property even though the purchaser intends to use the property for non-residential purposes, since it is suitable for use as a dwelling at the time of purchase.

b. Building treated by statute as use as a dwelling

2.4 A building used as residential accommodation for school pupils and other students, other than a hall of residence for students in higher or further education, is treated as a building used as a dwelling. This would apply, for example, to dormitory buildings in boarding schools, to blocks of flats let by universities, to students and student digs (s 116(2)).

2.5 Similarly, residential accommodation for members of the armed forces is a building used as a dwelling (s 116(2).

2.6 The same interpretation applies to an institution that is the sole or main residence of at least 90% of its residents (s 116(2)), but excluding a home or institution providing residential accommodation for children, or residential accommodation with personal care for others by reason of old age, or disablement, or alcohol or drug dependence, or mental disorder. The exclusions extend to a hospital or hospice, a prison, and an hotel (s 116(3)). An example of a property coming within the category would be a flat within a warden-assisted residential complex.

2.7 It is important to keep in mind the reason for needing to know what is residential and, hence, what then may be regarded as residential. There is no consistent approach to the meaning of residential in tax and other legislation. For example, in the case of value added tax (VAT), the treatment as a zero rated supply of the construction and disposal of new residential property has major financial consequences. In Group 5 of Schedule 8 to the Value Added Tax Act 1994, Note 4 defines various uses of buildings as being residential. The wording is the same as s 116 of FA 2003 but, apart from a hospital, a prison, and a hotel, what is not regarded as residential for SDLT is so regarded for VAT purposes. In this way, the same property enjoys favourable treatment under each tax regime. Note 4 also includes a monastery and a nunnery as residential for VAT and, while s 116 makes no reference to such uses, the definition giving examples of residential accommodation in the Statement of Practice SP1/04 *Stamp Duty Land Tax: Disadvantaged Area Relief* does include "accommodation for religious communities". The statement of practice, published by the Inland Revenue (now her Her Majesty's Revenue and Customs (HMRC)) in June 2004, examines, at length, the meaning of residential, and, in addition to comments on the statutory examples in paras 2.4 to 2.6 above, it makes specific comments on bed and breakfasts/guest houses. Occasional lettings of rooms, with no physical adaptations, will not change the status of residential, whereas a house occupied by the owners, who offer accommodation all year round as a business, means that the property is non-residential.

C. Building not used as but suitable for use as a dwelling

2.8 Section 116(1) includes within the definition of residential property "a building that is ... suitable for use as a dwelling". A guide to the interpretation of this phrase is contained in the Statement of Practice SP1/04.

2.9 In paras 15 to 18 of the statement, various examples are given of the HMRC view of buildings that are suitable for use as a dwelling. These include a house used as an office without particular adaptation. A building not in use, but last used as a dwelling, is also likely to be regarded as a dwelling. A further example is where two rooms of a house were used as a dentist's surgery and waiting room, when the whole property would normally be treated as a dwelling.

2.10 There are bound to be borderline cases as to whether a building is suitable for use as a dwelling. Given that the tax payable in a DDA, on any purchase price in excess of the £150,000 threshold, could be up to 4% of the purchase price if residential, or nil if not, the purchaser will, no doubt, argue strongly that the building is not suitable. The Statement of Practice states that attempts to render it unsuitable, such as removing a bathroom suite or kitchen facilities, do not, in the HMRC view, render it unsuitable. Even if the building is derelict or substantially altered, the taxpayer will need to provide evidence that it is no longer suitable. It is notoriously difficult to prove a negative.

2.11 This interpretation of the statute by HMRC is not, of course, binding. An alternative view is that s 116 refers to a building that is suitable for use. By adopting the present tense alone, this can be interpreted as meaning immediately suitable for use at the tax point. The section goes on to include, as residential, a building which "is in the process of being constructed or adapted" for use as a dwelling, again using the present tense. Nowhere does the section say "is or could be" or some similar wording. Thus, it is arguable that a building that is derelict or has no kitchen or bathroom is not suitable for use as a dwelling, since it requires work before it can be so used. Unless that work "is in the process of being" carried out, it means that the building could be made suitable for use, but is not suitable at present. Indeed, in para 34 of the Statement of Practice, HMRC give an example of an existing building which is being adapted for use, and the view of HMRC is that the building becomes residential when the work starts (or, if earlier, when the developer begins marketing the property). The example is "where a building is being made fit for habitation". If that is so, it must have been non-residential before the work commenced.

2.12 It seems certain that case law will have to emerge to provide tests as to the interpretation of "is suitable for use".

2.13 It is possible that HMRC view on attempts to render a building not suitable for residential use are derived from a need to prevent avoidance. If so, this would seem to be misplaced, as it is difficult to see why a vendor should damage a property to save a purchaser tax, while the purchaser would need to spend the tax saved on the restoration works.

d. Buildings being built for use as a dwelling

2.14 As stated in para 2.11, "a building that ... is in the process of being constructed" for use as a dwelling is residential. There is no statutory test of when the process begins. One approach would be to apply the time when a material operation is carried out in accordance with s 56 of the Town and County Planning Act 1990. The HMRC approach, in the Statement of Practice SP1/04 (see para 2.7), is "if a residential building is being built on it" at the time of transaction, which can be regarded as close to s 56 (see s 56(4)(a) "any work of construction in the course of the erection of a building").

e. Building being adapted for use as a dwelling

2.15 Adaptation for use as a dwelling involves the conversion of a non-residential property to residential use, such as converting a multi-storey warehouse to flats or lofts.

2.16 In this instance, where there is a change of use (and no external building works), the application of s 56 of the Town and County Planning Act 1990 (TCPA 1990) would provide for the building to become residential "at the time when the new use is instituted". This is in sharp contrast to the HMRC view, expressed in the Statement of Practice SP1/04, that the building becomes residential "when the developer begins marketing the properties for sale or starts physical work on the site, which ever is the earlier". This is to satisfy the terms of s 116 that "a building ... in the process of being ... adapted for such use" (as a dwelling) is a residential property. The Statement of Practice SP1/04 replaced the Statement of Practice SP1/03, which applied to stamp duty. The two statements of practice are very similar. However, SP1/04 does differ markedly in one regard. SP1/03 adopted as the test of when a building becomes residential as "when the builders start work". This is changed in SP1/04 to the test of when a developer starts work or begins marketing the property. The common aspect of when physical activity occurs does tie in with "in the process of being adapted" in s 116, but it is difficult to understand how marketing a property is a valid test in line with the section.

f. Buildings not in use but last used as a dwelling

2.17 A building may cease to be used as a dwelling and remain empty, often becoming in need of repair to restore it to a state for use as a dwelling to be resumed. For the purposes of SDLT, such a building ceases to be residential until it "is in the process of being ... adapted for such use".

2.18 In the case of bringing a derelict dwelling back into use, this may not involve any development in planning law terms, so that s 56 of the TCPA 1990 is inapplicable. This would be so if the building were still regarded as having a residential use before the work began. Again, HMRC's view is that the building becomes residential "when the developer begins marketing the properties for sale or starts physical work on the site which ever is the earlier".

g. Buildings partly in residential use

2.19 Section 116(6) provides that "building" includes part of a building. Hence, in a multi-use building, any part which is residential is treated in the same manner for SDLT as any building which is wholly residential, including the various situations referred to above.

2.20 There will normally be a single purchase price. If so, then there must be an apportionment of the price on such a basis as is just and reasonable between the residential and non-residential parts.

2.21 In the Statement of Practice SP1/04, HMRC accepts that apportionment is subjective, but gives examples of possible approaches based, seemingly, on areas. In most cases, such an approach is likely to be unreliable. Take, as an example, the purchase of a shop with a flat over in a good retail location, the floor space of each being the same. The likelihood is that the value of the shop greatly exceeds the value of the flat. In such circumstances a value apportionment will give a more just and reasonable apportionment. This will be the position in most purchases of mixed use buildings. Coincidently, it may result in less SDLT being payable.

h. Multiple purchases

2.22 It is recognised that the purchaser of residential property might do this in the course of a business as a property developer, property trader or property investor, which is no different from the purchase of commercial property, save for the use of the property.

2.23 Hence, s 116(7) provides that the purchase of the freehold or leasehold interest in six or more separate dwellings is not the purchase of residential property. The purchase must be by a single contract, even though completion might involve several instruments. The exemption applies even if the six or more dwellings are part of a larger portfolio being purchased, which includes non-residential property. In such a case the whole portfolio would be commercial.

2.24 A typical example would be the purchase of a block of six or more flats where the flats are self-contained though sharing common parts.

B. Residential land

i. Land forming part of the garden or grounds of a building

2.25 Section 116(1) includes, within the meaning of residential property, land that forms part of the garden or grounds of any residential building, as described above in section A, including any building or structure on the land. Typically, this would be the garden of a house, including any garage or shed or greenhouse or similar structures.

ii. Interaction with capital gains tax

2.26 In the Statement of Practice SP1/04, HMRC states that the test of whether land forms part of the garden or grounds of a residential building which it will apply is "similar to that applied for the purposes of the capital gains tax relief for main residences" in the Taxation of Chargeable Gains Act 1992 (s 222(3)).

2.27 The logic of this approach is understandable, although if a test is similar to another, this implies that it is not the same. The reason for this is that the meaning of residential property, dwelling-house and residence do not coincide in the capital gains tax (CGT) and SDLT legislation, and s 222 applies where the vendor occupies or has occupied the dwelling-house, which is not necessary for SDLT.

2.28 It is also the case that the application of s 222(3) in practice is far from straightforward.

2.29 Section 222(2) provides that land occupied with the residence as its garden or grounds up to 0.5 ha is included with the residence for exemption purposes. Section 222(3) provides for a larger area than 0.5 ha to be included if such a larger area is required for the reasonable enjoyment of the dwelling-house as a residence, having regard to the size and character of the dwelling-house. There have been many cases where s 222(3) has been considered, which is not surprising given the subjective nature of "required for the reasonable enjoyment", but various assumptions and tests have developed over the years. The cases are mainly unreported, apart from a few that are appeals against the commissioners' decision. The extent of the appropriate area of land is considered in IR *Tax Bulletin* 18 of 1995 and the *Capital Gains Tax Manual* published by HMRC.

2.30 Tax Bulletin 18 considers the HMRC view of the meaning of garden and grounds, and contains some helpful points. For CGT purposes there are three conditions, which will not all apply to SDLT.

1. It must be land which the owner has for occupation and enjoyment with the residence.
2. It must be the garden and grounds of the residence.
3. The area of land must not exceed the permitted area.

2.31 As mentioned above, SDLT requires the property to be residential in character, rather than owner occupied, to be subject to CGT relief. For SDLT purposes, it may be vacant, derelict or tenanted. The second and third conditions do apply. However, it must be appreciated that, if it can be demonstrated that a larger area than 0.5 ha is required for the reasonable enjoyment of the property, the permitted area can be greater. This is by no means an easy concept to prove.

2.32 While neither garden nor grounds are defined in statute, nor by any judicial authority, HMRC takes the view that a garden is an enclosed area devoted to the cultivation of flowers, fruit or vegetables. In relation to grounds, these are seen as "enclosed land surrounding or attached to a dwelling house or other building serving chiefly for ornament or recreation". The concept of enclosure does not necessarily mean a fence or physical barrier, as a property is enclosed by its boundaries of ownership.

2.33 Where land is used for agriculture, commercial woodlands, trade or business, it is not considered to be garden and grounds, neither is land fenced off for sale for redevelopment. However, land formerly part of the property's garden and grounds but unused or overgrown is. Buildings within the garden and grounds may form part of it if not used for a business purpose. Under certain circumstances, garden land may be severed from the residence, for example on the other side of a road, as some village gardens are. In some circumstances, such as the grounds of a guesthouse which is treated as residential, the land may be used partly for use of the guests and still be considered garden and grounds of the residence. The interaction between CGT legislation and SDLT legislation is likely to cause practical problems and should be approached with caution.

iii. Conflicts of interest for vendors and purchasers

2.34 It follows that the sale/purchase of a house standing in large grounds, only part of which is treated as required for the reasonable enjoyment of the house, will be a transaction involving residential and non-residential property.

2.35 From the vendor's point of view, if the sale comes within s 222, the vendor will argue for as large an area as possible, so as to minimise any CGT liability.

2.36 On the other hand, the purchaser will benefit from as much as possible being non-residential, so as to exploit the exempt band up to £150,000, or even more if the property is in a DDA (but see Chapter 18). However, the purchaser will one day be a vendor, and if the house is to be the purchaser's sole or main residence, a successful SDLT argument might prejudice the CGT argument at a later stage.

iv. Land being developed for residential use

2.37 Undeveloped land, not part of the garden or grounds of a dwelling-house, is non-residential even if it has planning permission to be developed for residential use. As explained in para 2.14 above, the view of HMRC is that it becomes residential "if a residential building is being built on it at the effective date of the transaction".

2.38 The importance of whether land in a DDA is residential or not is dealt with below. If the property is residential and the value is less than £150,000, it does not comprise chargeable consideration, and is therefore either, as residential not taxable, or if part of a mixed property, reduce the amount of tax payable.

2.39 However, where the development comprises six or more dwellings, the land remains non-residential — see para 2.23 above. The reason for this is that s 116(1) defines a residential property as including a building that is in the process of being constructed for use as a dwelling, which is taken to satisfy the meaning of "dwellings" in s 116(7), the sub-section that makes the provision for six or more dwellings being non-residential when purchased under a single contract.

C. Non-residential property

2.40 "Non-residential property means any property that is not residential property" (s 116(1) of FA 2003). This demonstrates the need to consider all possibilities of a property being taken to be residential before concluding that it is not.

D. Effect of Finance Act 2005 changes

2.41 The changes made by FA 2005 make the distinction between residential property and non-residential property of less importance than it previously was, apart from a purchase of a residential property not in a DDA in the price range £125,000 to £150,000. Where the advantages of being situated in a DDA are most applicable, in taxation terms, are where a mixed property is being sold. For example, a small hotel, situated in a coastal DDA, with a separate residential annex is sold for £630,000. The residential element which conveniently has its own council tax assessment, has a value of £135,000. This £135,000 does not count as chargeable consideration, therefore tax is payable at 3% on a value of £495,000, rather than 4% of £630,000. Difficulties may be experienced with HMRC in this area and it may be suggested that the tax payable should be based on 4% of £495,000. This wholly incorrect view is dealt with in Chapter 3. There are also variations in relation to leases and lease premiums for residential property situated in a DDA.

3 Designated Disadvantaged Areas

A. Origin of designated disadvantaged areas

3.1 Properties situated in designated disadvantaged areas (DDAs) attract disadvantaged areas relief (DAR), which is a measure which predates the introduction of stamp duty land tax (SDLT) in the Finance Act 2003 (FA 2003). The origins of DDAs came in a Government Urban White Paper *Our Towns and Cities: The Future: Delivering an Urban Renaissance* first published in November 2000. Initially the areas eligible for relief were designated enterprise areas in the pre-budget report of 2002.

3.2 DDAs are determined by applying the six weighted Indices of Deprivation, produced by Oxford University, covering matters such as income, employment, health and housing. Notwithstanding the deprivation tests, not all DDAs are the run down areas that might be anticipated.

3.3 DAR was originally provided for by s 92 of and Schedule 32 to the Finance Act 2001 (FA 2001), and was introduced on 30 November 2001. Initially, it was only intended to be available, as a partial relief, for conveyances or transfers on sale of both residential and non-residential (commercial) property for which the consideration did not exceed £150,000.

3.4 Stamp duty in respect of conveyances or transfers of commercial property in DDAs was abolished in consequence of the Stamp Duty (Disadvantaged Areas) (Application of Exemptions)

Regulations 2003 (SI 2003/1056) which had effect in relation to instruments executed on or after 10 April 2003. The relief in relation to residential property applies to purchases up to £150,000 (see Chapter 4). The Finance Act 2005 (FA 2005) returned the situation to the November 2001 position.

B. Property partly in a DDA

3.5 Where a property is partly in and partly outside a DDA, it falls to be apportioned on "a just and reasonable basis". However, if the property "outside" the DDA has the same postcode as that within, the whole will be treated as being within the DDA (see also para 3.10).

C. Ascertaining whether a property is within a DDA

3.6 In *Stamp Taxes Bulletin*, no 3, Her Majesty's Revenue and Customs (HMRC) introduced an interactive website whereby taxpayers could confirm whether a property falls within a qualifying area by reference to the property's postcode. The principle is that the taxpayer can go online to check if a postcode falls within a qualifying area, so as to be eligible wholly or in part for DAR. The postcode search tool can be found at *www.hmrc.gov.uk/so/dar/dar-search.htm*.

3.7 Where the taxpayer is unsure of the postcode, it can be found at: *www.consignia-online.com* or by telephoning The Royal Mail enquiry line on 08457 111222.

3.8 To perform a search, the taxpayer accesses the website at *www.hmrc.gov.uk/so* and clicks on "disadvantaged areas relief: new postcode search". At this point the taxpayer enters the relevant postcode, which must be in an exact postcode format, and the appropriate search will give either a positive or a negative result. Where a taxpayer believes that the property, in fact, falls within a DDA, even when a negative result has been returned, it may still be necessary to check whether or not the property is either newly built or has been subject to any ward boundary changes.

D. Current advantages of being located in a DDA

3.9 Residential and mixed property still enjoy certain advantages by being located within a DDA. These advantages are not always appreciated or understood by HMRC officials, although their own manuals are clear on the point. A residential property with a value of less than £150,000, situated in a DDA, is not subject to SDLT. Where a mixed property has a value, attributable to the residential portion of less than £150,000, this does not count as part of the chargeable consideration.

3.10 It has sometimes been suggested that SDLT incorporates a concept of total consideration, it does not. Consideration is either chargeable or it is not, therefore as per the example in Chapter 2, in a mixed transaction totalling £630,000, where the residential portion is say £135,000 the taxable consideration, against which the applicable rate of tax is judged, is £495,000. It is not 4% levied on £495,000, there is no statutory justification for this view.

3.11 The regulations with regard to lease premiums changed in the Finance Act 2008 (FA 2008), previously where the average rent exceeded £600 per annum (now amended to £1,000), any premium was taxed at the appropriate percentage for residential properties. This situation has been varied for commercial properties. However, for residential properties in a DDA, the SDLT payable on the rental element, which is no longer judged against an annual average, will be subject to the £150,000 threshold, the premium will be wholly taxable at the appropriate rate.

Tax Payable on Purchase

4

A. General

4.1 "The amount of tax chargeable in respect of a chargeable transaction is a percentage of the chargeable consideration for the transaction" (s 55(1) of Finance Act 2003 (FA 2003)). As to what constitutes a chargeable transaction is considered in Chapter 1. In this chapter, the determination of the chargeable consideration in respect of a purchase is explained. The meaning of chargeable consideration is set out in s 50 of and Schedule 4 to the FA 2003.

B. Chargeable consideration

i. Single payment

4.2 In the majority of cases the purchase of an interest in a property is the payment of one agreed amount, which is the chargeable consideration. It may be that a part payment is made on exchange of contracts, but that is part of the agreed amount. Stamp duty land tax (SDLT) is payable on the agreed amount. The time of payment where completion is preceded by a contract is considered in paras 1.7 to 1.17.

4.3 As stated, the agreed amount paid determines the tax payable, so that, for example, where the transaction is a gift with no consideration, there is no SDLT payable (Schedule 3 to FA 2003, para 1) (although market value may be substituted for the operation of other taxes). The one exception is where the purchaser is a company purchasing from a connected vendor, when market value will be adopted if different from the actual consideration (s 53 of FA 2003, and see Chapter 10 and also para 4.23).

4.4 The position is the same where payment is made by someone connected with the company purchaser (see Chapter 16 for the meaning of "connected").

4.5 If the price paid includes value added tax (VAT) because, for example, it is a new non-residential building, or the vendor has opted to tax, SDLT is payable on the VAT inclusive price, but not otherwise. Unlike stamp duty before, if no VAT is included but the price could be subject to VAT after the effective date of the transaction, by virtue of a later election to opt to tax, the potential VAT payment is left out of account (Schedule 4 to FA 2003, para 2). (For a detailed consideration of Stamp Duty Land Tax (SDLT) and VAT see Chapter 8.)

4.6 In the case of collective enfranchisement when an Right to Enfranchise (RTE) company purchases the landlord's interest, a single price will be payable. The price is then divided by the number of flats, and that fraction of the total consideration determines the rate of tax which is then applied to the total chargeable consideration (s 74 of FA 2003, and see Example 4E).

4.7 The right of collective enfranchisement is derived from Part 1 of the Landlord and Tenant Act 1987 or Chapter 1 of Part 1 of the Leasehold Reform, Housing and Urban Development Act 1993. An RTE company is the vehicle for the several leaseholders exercising the right of collective enfranchisement to acquire the freehold interest.

4.8 A similar approach is adopted when two or more crofts are purchased under the crafting community right to buy powers derived from Part 3 of the Land Reform (Scotland) Act 2003, the total consideration being divided by the number of crofts being purchased (s 75).

4.9 If a single price is agreed for the purchase of two or more properties, or for the purchase of a property together with some other matter, the price is to be apportioned on a just and reasonable basis. In most cases, this is likely to be a value apportionment (Schedule 4, para 4).

4.10 On the other hand, if there is a single bargain, but separate consideration is given or purports to be given for various elements (which would consequently reduce the tax payable), tax is payable on a "just and reasonable" apportionment of the aggregate consideration (Schedule 4, para 4 of the FA 2003). This issue is dealt with in more detail in Chapter 10, originally aimed at the problem of purchasers, by agreement with vendors, attributing unreasonable amounts to non-taxable items such as chattels, its application has spread into the area of composite transactions where a property and the business carried on therein are sold together. Elements such as "goodwill", where it comprises an intangible asset, are not taxable to SDLT and therefore a "just and reasonable" apportionment of the component parts of the transaction will be required.

ii. Payment other than cash

4.11 In those cases where the consideration is part in money and part in some other form, or indeed wholly in another form (money's worth rather than money), the market value of the non-money consideration at the effective date of the transaction is the value of the consideration (Schedule 4, para 7 of the FA 2003, and see Chapter 10 for meaning of "market value"). Where the non-money consideration is the provision of services, the value is the amount that would have to be paid in the open market to obtain those services (Schedule 4, para 11).

4.12 For example, if a residential developer purchases land for development and agrees, in addition to paying a sum of money for the land, to sell back one of the houses to be built to the vendor at cost, the market value of buying at cost is added to the price of the land for SDLT purposes. Say the market value of the house is £300,000 and the vendor will have to pay only £160,000 to buy it, a figure around £140,000 would be added to the money paid by the developer for the land. The vendor, when buying back the new house, will pay SDLT on £160,000.

4.13 Where a transaction includes a requirement for the vendor or someone connected to the vendor to carry out work, the value of such work will be added to any other money payment by the purchaser to determine the chargeable consideration. The value of the work is the sum that would have to be paid in the open market to have the work carried out (Schedule 4, para 10).

4.14 This is principally an anti-avoidance measure to stop the loss of tax by the scheme whereby a developer sells some land, typically a plot of land for the building of a house, on the basis that the developer will build the house for the purchaser. In this way, the tax payable would be on the price of the land only (perhaps less than its market value), in the absence of this anti-avoidance measure.

4.15 The measure does not apply where the works are carried out after the effective date of the transaction and are to be carried out on the land the subject of the transaction, or other land owned by the purchaser or someone connected to the purchaser, and it is not a condition of the transaction that the works are to be carried out by the vendor or someone connected to the vendor.

4.16 In a press release in April 2004 (*SDLT — sale of land with associated construction, & c, contract*), the Inland Revenue (now Her Majesty's Revenue and Customs or (HMRC)) stated that "we have been asked how to determine the chargeable consideration for SDLT purposes where V agrees to sell land to P and V also agrees to carry out works (commonly works of construction, improvement or repair) on the land sold. Our view is that the decision in *Prudential Assurance Co Ltd* v *IRC* [1992] STC 863 applies for the purposes of SDLT as it does for stamp duty. This is because the basis of the decision was the identification of the subject-matter of the transaction and this is as relevant for SDLT as it is for stamp duty". Where the sale of land and construction contract are, in substance, one bargain then, following *Prudential*, there must be a just and reasonable apportionment.

iii. Phased payments

4.17 Where all or part of the consideration is payable at an agreed date after the transaction, no discount is given for such postponed payments (Schedule 4, para 3).

4.18 For example, if some development land is purchased under a single contract as a single transaction and payment is agreed to be spread over three years, by four equal payments of £500,000, one at the time of the transaction and three further payments at the end of years one, two and three, SDLT will be payable on 4 × £500,000, equal to £2 million, even though the value to the vendor, and the cost to the purchaser, is less than £2 million.

iv. Contingent payments

4.19 Some transactions are structured to provide for further payments to the vendor contingent on a future event taking place. The treatment of the consideration in such cases is set out in s 51 of FA 2003. Typically, a payment is made at the time of the transaction, with provision for a further payment on, say, receipt of planning permission. The further payment may be as fixed in the agreement leading to the transaction, or may be a figure to be determined at the time of the contingent event. For example, a farmer sells land to a developer for £100,000 with a further payment to be made of £900,000 on the grant of planning permission for residential development.

4.20 However, under s 90 of FA 2003, if the further payment will be made, or may be made, more than six months after the effective date of the transaction, the purchaser may apply to HMRC to defer payment of tax on the contingent sum. The position when the contingent event is realised is considered in Chapter 5.

4.21 Alternatively, the farmer sells the land to a developer for £100,000 with a further payment to be made of 90% of the market value of the land if planning permission is granted, at the time it is granted. In such a case "a reasonable estimate" of the further payment must be made and added to the payment of £100,000, the whole being treated as chargeable. It is likely that a purchaser, in such circumstances, might seek advice from a valuer, who would need to check whether offering estimates of future values is covered by their professional insurance. This type of case is one where consideration deferment under s 90 is likely to be applied, as it depends on a future contingent event to trigger the additional payment.

v. Debt as consideration

4.22 In some transactions, the purchaser may include in the terms an allowance involving debt, which is reflected in the overall consideration. For example, the purchaser may release the vendor from a debt owed to the purchaser, or may take over responsibility for a debt owed by the vendor, such as taking over the vendor's outstanding mortgage.

4.23 In such circumstances, the amount of debt released or assumed is part of the chargeable consideration (Schedule 4, para 8 of the

FA 2003). However, if the addition of this outstanding debt takes the chargeable consideration above market value, the chargeable consideration is limited to the market value.

4.24 Where the debt is secured on the property which is the subject matter of the transaction, and two or more persons hold an undivided share in the property before or after the transaction, the debt is apportioned in the proportion of the undivided share to the whole property. Joint tenants are treated as holding equal undivided shares of the property. For example, A pays B £200,000 for a half share interest in a property and at the same time assumes joint responsibiliy for the outstanding mortgage of £300,000. Although A is jointly and severally liable (with B) for the whole of the debt for SDLT purposes, they are deemed to have taken on responsibility for 50% thereof, the consideration is therefore deemed to be £450,000.

vi. Exchanges

4.25 When a transaction involves the sale of a property where the vendor also purchases a property from the purchaser, the chargeable consideration in respect of each purchase is the market value of the property being purchased (Schedule 4, para 5 of the FA 2003). So, if A sells Blackacre to B in return for the purchase of Greenacre from B, the chargeable consideration on which B pays SDLT for the purchase of Blackacre is its market value, while A pays SDLT on the market value of Greenacre.

4.26 Prior to the Finance Act 2007 (FA 2007), where A and B were connected persons, then the consideration was aggregated, on the basis that this was a linked transaction (s 108) and the rate of tax applicable was based on the aggregated figure, although the chargeable consideration for each party was taken to be the market value as in para 4.25 above. This unfair provision has now been dropped.

4.27 This applies where the subject matter of the sale or purchase is a major interest in land (basically a freehold or leasehold interest or its equivalent in Scotland). If none of the transactions is in respect of a major interest, the chargeable consideration is the amount or value of any consideration given for the acquisition other than the disposal. If the acquisition is the grant of a lease, at a rent, the chargeable consideration is that rent.

4.28 In certain instances, where a house building company acquires from the purchaser of one of its new houses the interest in the house owned by the purchaser, the company is not liable to pay SDLT on its purchase. This is considered further in Chapter 1.

vii. Linked transactions

4.29 If a transaction forms part of a number of linked transactions, then the consideration is the aggregated total. This is fundamentally an anti-avoidance mechanism. Transactions are "linked if they form part of a single scheme, arrangement or series of transactions between the same vendor and purchaser or, in either case, persons connected with them" (s 108 of the FA 2003). The Income and Corporation Taxes Act 1988 (ICTA 1988) (see Chapter 16) gives the meaning of "connected" in s 839. There are specific rules for leases in this regard, which are dealt with in Chapter 6.

4.30 There are provisions in s 108, similar to those in s 103 (joint purchasers) for persons involved in linked transactions to use a single form of return where the transactions occur on the same date.

4.31 The attitude of HMRC as to what comprises a linked transaction appears to have changed, although the SDLT manuals have not. Say, for example, that a purchaser buys three flats in a new development. If the purchase attracts a discount or a variation of terms reflecting the "block" purchase, this will be seen as a linked transaction. However, if the three flats were purchased in the same manner, and for the same value, as if they had been separate transactions, then it appears that this will not be seen as linked. This is a different approach to that was initially experienced in practice, at one point, if say a husband and wife each purchased a single flat as an investment, in the same block, from the same purchaser, then this was apparently deemed a linked transaction. The view now taken is if the transaction does not show some "variation from the norm", it is not deemed to be a linked transaction.

4.32 There are exceptions from the linked transactions rule in certain collective enfranchisement or crofter's Right to Buy (RTB) transactions.

viii. Payments for chattels

4.33 It is not uncommon for items other than the property itself to be included in a transaction, for example furniture, carpets and curtains with a house, or plant and machinery, and fixtures and fittings with a commercial property. Since SDLT is payable in respect of a chargeable land transaction, it follows that items other than the property itself should be left out of account.

4.34 Since the payment structure of SDLT is illogical, throwing up high marginal rates of tax as the consideration moves from one band to another (see para 4.39 and Examples 4C and 4D), there is a temptation to inflate the price paid for chattels so as to keep the consideration for the property itself in a lower band. It is impossible to say how widespread is this practice, but HMRC have made it clear that they intend to attempt to stop it. The provisions of para 4 to Schedule 4 of FA 2003 allow for the aggregation and re-apportionment of composite transactions on a "just and reasonable" basis, primarily to prevent this form of tax avoidance.

4.35 It is useful to consider what comprise, in HMRC's view, chattels, fixtures and fittings. Their view may not be either definitive or correct, in each case however it is as follows, items which are considered chattels (mainly in a residential context) normally include:

- carpets, fitted or otherwise
- curtains and blinds
- free standing furniture
- kitchen white goods
- electric and gas fires where removable after disconnection from the supply without damage to the item or property
- light shades and fittings unless recessed.

Items normally regarded as part of the property include:

- fitted kitchen units, cupboards and sinks
- agas and wall mounted ovens
- fitted bathroom sanitary ware
- central heating systems
- alarm systems
- plants, shrubs, etc except in pots.

4.36 Whether plant and machinery is a fixture or a chattel is determined in the same way as for any other asset and a similar approach is adopted as in the residential context above. Where a tenant has the right to sever a fixture this may be a chargeable interest (s 48 (1) FA 2003). If an item can be removed intact without damage to the item or the property, in the commercial context, then it will probably be a fixture, however, if it is an integral part of the structure, for example an escalator, it will be plant and part of the property.

ix. Annuities

4.37 Where the chargeable consideration consists of an annuity payable for life, in perpetuity, for an indefinite period, or for a definite period exceeding 12 years, the consideration is limited to 12 annual payments, starting from the date of the transaction (s 52 of FA 2003). If payments vary, the 12 highest payments are adopted, but adjustments for changes in the retail price index are ignored. The consideration is the aggregate undiscounted amount, as for phased payments (see para 4.17). The provisions if payments are contingent, uncertain or unascertained are the same as those set out in paras 4.19 to 4.21.

C. Amount of tax payable — residential

i. Not within a designated disadvantaged area

4.38 Property treated as being residential is examined in Chapter 2. A designated disadvantaged area (DDA) is defined and explained in Chapter 3.

4.39 The amount of SDLT payable on the purchase of a residential property not within a DDA (s 55 of FA 2003) is:

Consideration		SDLT payable on consideration
	Up to £125,000	Nil
Over £125,000	Up to £250,000	1%
Over £250,000	Up to £500,000	3%
Over £500,000		4%

The Finance Act 2007 raised the Nil band to £125,000 for purchases after 16 March 2007 raised for one year to £175,000 until September 2009.

Example 4A

X purchases a residential property for £240,000. SDLT payable @ 1% = £2,400.

Note that the whole consideration is subject to SDLT, not that in excess of £125,000 (which would be logical and fair).

Example 4B

X purchases a residential property for £250,000. SDLT payable @ 1% = £2,500.

Example 4C

X purchases a residential property for £251,000. SDLT payable @ 3% = £7,530.

Note that an additional payment of £5,030 is due on the increase in the consideration of £1,000 — a marginal rate of 503% (which is illogical and unfair).

Example 4D

X purchases a residential property for £260,000 including £10,000 for curtains and carpets. SDLT payable at 1% of £250,000 = £2,500.

Note that if HMRC successfully argue that the true value of the chattels was less, say £9,000, then SDLT payable on £251,000 = SDLT payable of £7,530 as in Example 4C.

Example 4A

The owners of four flats purchase the freehold interest for £550,000 under the right of collective enfranchisement (see para 4.6). Dividing the consideration (£550,000) by the number of flats (four) produces £137,500.

SDLT rate of £137,500 is 1%

Hence, SDLT payable is 1% of £550,000.

ii. Within a DDA

4.40 SDLT payable on the purchase of a residential property within a DDA is the same as on the purchase of a property not within a DDA, save that the nil rate band is up to £150,000, and the 1% band is consequently over £150,000 up to £250,000 (Schedule 6, para 5 of FA 2003).

D. Amount of tax payable — non residential

i. Not within a DDA

4.41 The amount of tax payable on the purchase of a non-residential property not within a DDA (s 55) is:

Consideration		SDLT payable on consideration
	Up to £150,000	Nil
Over £150,000	Up to £250,000	1%
Over £250,000	Up to £500,000	3%
Over £500,000		4%

4.42 It is clear that, for a purchase between £60,000 or £120,000 up to £150,000, it will be important to establish whether a property is non-residential, with no tax to pay, or is residential with tax at 1% but see para 4.31. As the one year uplift to the residential property threshold does not apply to DDAs, for a purchase between £150,000 and £175,000, it is important to know whether a residential property is in a DDA with tax to pay at 1%, or outside a DDA with none to pay up to £175,000.

ii. Within a DDA

4.43 No tax was payable on the purchase before 17 March 2005 of non-residential property within a DDA (Schedule 6, para 4 of FA 2003). It was even more important to establish whether a property was non-residential or residential on purchases over £150,000, where the difference could have been as much as 4% on purchase prices over £500,000. For purchases after 16 March 2005, the position is the same as property not within a DDA (see para 4.34 and Chapter 18).

5 Tax Payable or Repayable after Purchase

A. Resolution of contingent event

5.1 In Chapter 4 (paras 4.19 to 4.21), the treatment of transactions where a further payment may be payable contingent on a future event was examined. The further payment may be one fixed at the time of the transaction, or a sum to be assessed when the contingent event is realised, such as a percentage of the prevailing market value.

i. Further payment not deferred by s 90

5.2 If payment of stamp duty land tax (SDLT) was not deferred under s 90 of Finance Act 2003 (FA 2003), then an adjustment may be required of the SDLT payable once the contingent event is realised, or it becomes clear that it will not occur. For example, where a further payment is contingent on the grant of planning permission, the contingent event is realised if permission is granted. On the other hand, it becomes clear it will not occur if permission is refused, perhaps after passing through the appeal procedures.

5.3 Provisions for determining the tax position are contained in s 80 of FA 2003. If the contingent event is realised, the amount of consideration that becomes payable as a result is to be calculated. If the consideration is greater than anticipated at the time of the original transaction, so that further SDLT becomes payable, a further return must be made and the additional tax must be paid within 30 days of the contingent event being realised (see Chapter 13, paras 13.25 to 13.28). The additional tax is calculated by applying the rates applicable at the effective date of the transaction, which is "the date of the event as a result of which the return is required" (s 80(3) of FA 2003).

Example 5A

X purchases land from a farmer in 2004 and pays £100,000. A further payment, based on the market value of the land, will be made if planning permission is granted for residential development.

X estimates that the further payment will be £700,000. Hence, SDLT is payable at 4% of £800,000 = £32,000.

In 2007, planning permission is granted. Land values have risen so that a further payment of £850,000 is to be made. The total consideration is thus £950,000. Hence, SDLT payable (assuming no change in the rate of tax) on actual consideration is:

4% of £950,000	£38,000
Deduct SDLT paid at time of transaction	£32,000
Additional SDLT payable	£6,000

5.4 In addition to the further payment of tax, interest is payable on the additional SDLT payable from the end of 30 days after the later effective date of the transaction to the date of payment of the additional SDLT (s 87 of FA 2003). The additional SDLT is payable (and is likely to be paid) within 30 days.

5.5 If, on the other hand, it becomes clear that the contingent event will not occur, any overpaid SDLT is refunded, together with interest from the time of payment to the time of the refund. Interest is not income for tax purposes (s 89 of FA 2003).

Example 5B

X purchases land from a farmer in 2004 and pays £100,000. A further payment of £900,000 will be made on the grant of planning permission for a specified development.

Hence, SDLT is payable at 4% of £1,000,000 = £40,000.

In 2007, planning permission is refused and it is clear that any appeal will be fruitless. Hence, SDLT payable is limited to the consideration of £100,000, which is nil.

X will recover the SDLT paid of £40,000 plus interest from when the tax was paid on the purchase in 2004 to the date when the order for repayment is issued in 2007.

5.6 The provisions regarding the resolution of a contingent event apply equally to the ascertainment of what was an unascertained or uncertain consideration. As SDLT becomes payable at a specific point in time, subsequent falls in value do not trigger a right to repayment.

ii. Further payment deferred by s 90

5.7 Where a transaction involves further possible consideration related to a future contingent event, and the payment of SDLT on that further consideration is deferred under s 90 of FA 2003, payment of SDLT will be triggered when the contingent event is realised.

5.8 As before, the tax payable on the consideration is to be assessed adopting rates applicable at the time of the transaction and the time of the event triggering the deferred payment, and a return with payment must be submitted within 30 days. Interest is payable as it is in the case where payment was not deferred under s 90, if payment of SDLT payable is not made within 30 days of the date when the deferred payment of consideration is due.

5.9 If it becomes clear that the contingent event will not occur, then no further action is required.

6 Tax Payable on Start of Lease

A. General

6.1 When taking a new lease, the lessee will be liable to pay stamp duty land tax (SDLT) on the consideration payable. In England, Wales and Northern Ireland, a lease means "an interest or right in or over land for a term of years (whether fixed or periodic)" (s 120 of Finance Act 2003 (FA 2003) and Schedule 17A to FA 2003, para 1). It also means "a tenancy at will or other interest or right in or over land terminable by notice at any time" which is curious given that a tenancy at will and a licence to use or occupy land are not chargeable interests (see para 1.5). A lease in Scotland has its usual meaning (Schedule 17A, para 19).

6.2 In most cases, the consideration will include rent payable, although in some cases leases are taken where the rent is nil — a peppercorn if demanded.

6.3 In addition to rent, the lessee may pay a premium, which is a capital sum payable to the lessor, or there may be two or more sums payable at specified times in the future.

6.4 The calculation of SDLT payable on a premium or premiums is the same as that for other capital payments, as described in Chapter 4. Any capital payment by a lessor to a lessee for taking the lease, a reverse premium, is not subject to SDLT (Schedule 17A to FA 2003, para 18). As regards rent, SDLT is payable on the capital equivalent of the rents, calculated as the discounted amount, or net present value (NPV). The calculation is made in accordance with s 56 of and Schedule 5 to FA 2003. The taxable amount is the NPV in the case of residential property, £125,000, and of £150,000 in other cases. The rate of tax is 1%.

B. Calculation of net present value

i. Discount rate

6.5 The present value of the right to receive a sum in the future is determined by discounting the sum for the waiting period at the appropriate discount rate (the temporal discount rate) (Schedule 5 to FA 2003, para 8). This is a process familiar to valuers who adopt this approach in valuing future receipts and income flows when applying the investment method of valuation or a discounted cash flow (DCF) method.

6.6 In the case of SDLT, the discount rate is one determined by the Treasury. The current rate is 3.5%. In terms of property investments, this yield is less than nearly every property yield, in many cases significantly less. Consequently, the NPV of the rent payable, for SDLT purposes, exceeds the market capital value (the price) which a purchaser would pay for the interest which would entitle the owner thereof to receive those rents.

ii. Rent payable

6.7 The rent payable, which is to be discounted, is the rent reserved under the lease. Where the rent is inclusive of other matters, such as insurance, or rates, or repairs, without apportionment, the whole sum is treated as rent (Schedule 17A to FA 2003, para 6). However, this is without prejudice to the provision for a just and reasonable apportionment under Schedule 4, para 4. In any event, the tenant's obligations for such matters as to repair, to insure, to pay service charges and other non-rent matters, are not taken to be chargeable consideration (Schedule 17A, para 10). It is in the tenant's interest for the rent to be stated separately from any other payment.

iii. Net present value

6.8 The calculation of the NPV (V) is made by applying the formula:

$$v = \sum_{i=1}^{n} \frac{r_i}{(1 + T)^i}$$

where:
r_i is the rent payable in year i
i is the first, second, third, etc year of the term
n is the term of the lease, and
t is the temporal discount rate

in para 3 of Schedule 5 to FA 2003. The NPV is the relevant rental value for the calculation of the tax payable.

6.9 The calculation is the same as the sum of the present values of £1 per annum or Years Purchase single rate, at 3.5%. These are contained in AW Davidson, *Parry's Valuation & Investment Tables*, 12th ed (Estates Gazette, 2002). The relevant extract is contained in Appendix 2.

6.10 It is surprising to some that the Her Majesty's Revenue and Customs (HMRC) should adopt a calculation which discounts the rent payable during a year at the end of that year, rather than at the time of actual payment, commonly quarterly in advance. This perpetuates the debate among valuers and others as to the validity of valuing rental income assuming it is paid annually in arrear, the traditional method, against valuing the income when paid, usually quarterly in advance.

6.11 Apart from applying the figures from the YP Single Rate Table, as is done in the examples below, the figure can be obtained from *Parry's Valuation & Investment Tables*. It was formerly included in the HMRC website in the Press Release *SDLT — Guidance notes on self-certification* issued by HMRC in November 2003 which also contains the present value of £1 for each year up to 99 years to eight places of decimals compared with seven places in *Parry's Valuation & Investment Tables*. However, there was an error in the present value tables after year 6 when the notes were first published and it has been withdrawn.

C. Rent subject to future contingent event

6.12 Where the rent is subject to variation dependent on a future contingent event, the position is as described in Chapter 4 (see paras 4.19 to 4.21). However, the possibility of deferring payment until the outcome of the contingent event is known, under s 90 of FA 2003, is not available (s 90(7)).

6.13 So, for example, if a lease of a warehouse is granted at a rent reflecting its use as a warehouse, but with a provision for a higher rent to be paid if planning permission is granted for retail use, SDLT will be assessed on the higher rent as stated, or a reasonable estimate of the rent payable if not stated. It would seem appropriate to include an estimate of the time before the outcome of the contingent event will be known, and to incorporate that delay in the calculation of the NPV. Once the outcome is known, then adjustments for further payments or refunds of SDLT will be made as appropriate.

6.14 A provision for rent to be varied in line with changes in the retail price index is not treated as a rent subject to variation dependent on a future contingent event (Schedule 17A to FA 2003, para 7).

D. Rent reviews

6.15 Where a rent is to be reviewed to reflect prevailing market rental values at the time of the review, any review which comes into effect up to the end of the first five years of the term of the lease is incorporated in the calculation of the NPV (Schedule 17A, para 7 of FA 2003, and see Example 6.1 below). This requires the lessee to predict the changes in rental value over the period up to the review date, presumably on the advice of a valuer. The forecast rent is then incorporated in the NPV calculation. If, at review, the rent is higher than the forecast rent then increased SDLT becomes payable, calculated by re-computing the original NPV but adopting the now known reviewed rent, additional SDLT becoming payable on the increase in the NPV. If lower, the same process is followed, with a refund of SDLT on the reduction in the NPV.

6.16 In some cases, a lessee takes a lease where a rent review is stated as falling five years from a date which is an earlier date than the commencement date of the lease ("a specified date"), perhaps the preceding quarter day. If this specified date is within the three months preceding the commencement date, and there is a rent review five years after the specified date, it is assumed that the review date is five years after of the beginning of the term (Schedule 17A, para 7A). This removes it from the need to forecast rent as seen in para 6.15.

6.17 Rent reviews which are effective after five years or more are not taken into account in determining NPV (but see Chapter 7, section C below).

Example 6.1 Lease, only rent payable, rent review within first five years

A residential property is let for 12 years at a rent of £10,000 per annum (pa), subject to upward only rent reviews after four and eight years.

Initial rent			£10,000 pa
Predicted rent after 4 years — assume rental value increase of 3% pa			
Initial rent			£10,000 pa
Amount of £1 for 4 yrs @ 3%			1.1255
Predicted rent			£11,255 pa
Calculate NPV			
Initial rent		£10,000 pa	
YP 4 yrs @ 3.5%	3.6731	£36,731	
Rent after 4 yrs	£11,255 pa		
YP 12 yrs @ 3.5%	9.6633		
Less YP 4 yrs @ 3.5%	3.6731	5.9902	£67,420
NPV			£104,151
Less exempt amount			£60,000
SDLT @ 1% on			£44,151
SDLT payable			£441

Note: The rent review after five years is ignored. The calculation will be repeated when the rent is fixed at the first review (see para 7.1). A letting which takes place after 16 March 2005 has an exempt amount of £125,000.

E. Term of lease

i. A fixed term

6.18 The length of the term of a lease is a key factor in calculating the NPV. Where the lease is for a fixed term, the term is taken as the contractual term specified in the lease or, if shorter, the period from the date of the grant of the lease until the end of the contractual term. The right of any party to determine or renew a lease, such as a break clause or an option to extend, is ignored in determining the term (Schedule 17A, para 2).

Example 6.2 Lease, only rent payable

A commercial property is let for 10 years at a rent of £100,000 pa

Rent	£100,000	pa
YP 10 yrs @ 3.5%	8.3166	
NPV	£831,660	
Less exempt amount	£150,000	
SDLT @ 1% on	£681,660	
SDLT payable	£6,816	

Example 6.3 Lease, only rent payable after rent free period

A commercial property is let for 10 years at a rent of £100,000 pa, with an initial rent free period of six months.

Rent	Year 1		£50,000	
	YP 1 yr @ 3.5%		0.9662	£48,310
Rent	Years 2–10		£100,000 pa	
	YP 10 yrs @ 3.5%	8.3166		
	Less YP 1 yr @ 3.5%	0.9662	7,3504	£735,040
NPV				£783,350
Less exempt amount				£150,000
SDLT@ 1% on				£633,350
SDLT payable				£6,333

Example 6.4 Lease, rent payable plus initial premium

A flat is let for 99 years subject to a premium payable of £180,000 and ground rent of £100 pa which doubles after 33 years and 66 years.

Rent	£100	pa
YP 99 yrs @ 3.5%	27.6234	
NPV	£2,762	

Note: The increased rent payable after five years is ignored. The NPV will be taxed separately from the premium to determine the SDLT payable (in this case nil as below the threshold of £125,000, and see sections H and I below).

6.19 As an option to extend a lease is ignored in determining the length of the term, this allows for some tax mitigation. If the lessee wishes to take a lease for, say, 20 years, the initial payment of SDLT will be reduced if the lease is for a term of, say, 12 years, with an option to extend the lease (or to take a new lease) for eight years. The landlord would probably require cross options. It might be possible for the initial lease to be for a term where the rent has a NPV below the exempt limit. The second payment may be higher if rents increase, but it is unlikely that the lessee would be worse off in discounted cash flow terms.

ii. Renewal of fixed term lease

6.20 The renewal of a fixed term lease is treated as a new lease commencing on the expiry of the previous lease. This applies in particular to business leases granted under Part 2 of the Landlord and Tenant Act 1954 or leases under the Business Tenancies (Northern Ireland) Order 1996 (SI 1996/725 (NIS)). However, if a fixed term lease is not brought to an end, so that the lessee remains in possession, the lease is treated as being extended (tacit relocation in Scotland).

6.21 For SDLT purposes, the lessee is treated as taking a new lease from the start of the fixed term lease for a period of 12 months longer than the original lease. If the situation has not changed after 12 months, it is assumed that a further lease for 12 months has been taken, and so on until a new fixed term lease is taken (Schedule 17A, para 3). The 12-month extensions will increase the NPV so that if, initially, they are within an exempt level NPV, the cumulative adjustments could bring the lessee into a taxable NPV.

6.22 If, at the end of the fixed term, the lessee is liable to pay an interim rent, but this has not been agreed or may not be known until determined by the court, the provision relating to contingent events will apply (see paras 6.11 and 6.12 and Example 6.6 below).

Example 6.5 Lessee remains in possession after expiry of fixed term

Lessee occupies shop under a lease for a term of five years which has expired. Lessee continues to pay rent of £30,000 pa which landlord accepts.

Assume lease was granted for six years.

Rent	£30,000 pa
YP 6 yrs @ 3.5%	5.3286
NPV	£159,858

Compare original position	
Rent	£30,000 pa
YP 5 yrs @ 3.5%	4.5151
NPV	£135,453
Increase in NPV	£12,203
(and now tax payable on excess above £150,000)	

Example 6.6 as for Example 6.5, but landlord has applied for an interim rent

Lessee occupies shop as for Example 6.5, paying rent of £30,000 pa. Landlord has applied to court for an interim rent under Landlord and Tenant Act 1954.

Predict Interim Rent	
Estimated current rental value	£40,000 pa
Rent payable at end of lease	£30,000 pa
Interim rent, predicted	£35,000 pa

Calculate NPV			
Rent		£30,000 pa	
YP 5 yrs @ 3.5%		4.5151	£135,453
Reversion to:		£35,000 pa	
YP 6 yrs @ 3.5%	5.3286		
Less YP 5 yrs @ 3.5%	4.5151	0.8135	£28,473
NPV			£163,926

Note: If interim rent is not determined at £20,000, then NPV will be recalculated, with consequent adjustment to the tax liability of the lessee.

iii. *Agreement for lease preceding a lease*

6.23 An agreement for a lease is commonly entered into where the lessee is to carry out development. The agreement provides that the lessee will carry out specified work, which can range from alterations to an existing building to the construction of a major development. On satisfactory completion of the work, the lease attached to the agreement comes into effect.

6.24 For the purposes of SDLT, the term of the lease commences from the date of substantial performance of the agreement and ends on the contractual termination date of the lease. A contract will be substantially performed when the lessee takes possession of the property under the terms of the agreement for a lease. When the lease is subsequently granted, it is treated as a surrender and re-grant for the variation of the term of a lease (Schedule 17A, para 12A, and see para 6.30 below).

iv. Lease for indefinite term

6.25 Where a lease is granted for an indefinite term, it is treated as a lease for 12 months, in the same way as a lessee who remains in occupation after the expiry of a fixed term, as described in para 6.21. Examples given in Schedule 17A to FA 2003, para 4, include a periodic tenancy or other interest terminable by a period of notice, a tenancy at will, or any other interest terminable by notice at any time.

F. Variation of a lease

i. Variation of rent

6.26 Where a lease is varied so as to increase the amount of rent payable, the variation is treated as the grant of a new lease, the rent payable being the increase in the rent (see Example 6.7). This does not apply to an increase in rent already provided for in the lease (Schedule 17A to FA 2003, para 13) nor to a lease where stamp duty was paid — a lease commencing before 1 December 2003 (Schedule 17A, para 9(4)).

6.27 Where a lease is varied so as to reduce the amount of rent payable, the variation is treated as an acquisition of a chargeable interest by the lessee (Schedule 17A, para 15A(1)). However, if no payment is made by the lessee for the variation, then there will be no consideration and therefore no tax payable.

Example 6.7 Variation of rent payable under a lease

A lease of a warehouse was granted for 15 years at a rent of £120,000 per annum (pa) without review. The lease has eight years to run. The lease is to be varied by allowing non-food retail use and the lessee has agreed to pay a revised rent of £200,000 pa.

Rent payable	£200,000	pa
Less old rent	£120,000	pa
Additional rent	£80,000	pa
YP 8 yrs @ 3.5%	6.8740	
NPV	£549,920	

Less exempt amount	£150,000
SDLT@ 1% on	£399,920
SDLT payable	£3,999

ii. Variation of term

6.28 Where the period of a lease is extended, this is treated as the grant of a new lease for the extended term. This new lease is treated as a transaction which is linked to the original grant.

6.29 Where the period of a lease is reduced, this is treated as the acquisition of a chargeable interest by the lessor (Schedule 17A to the FA 2003, para 15A(2)). As in the case of the reduction in the amount of rent payable, if the lessor receives no consideration for the reduction in the term, then there is no consideration on which tax will be payable.

Example 6.8 Variation of both rent payable under a lease and of the term of the lease

As for Example 6.7, but lease will be extended by a further seven years.

Next 8 years		
Rent payable	£200,000	pa
Less old rent	£120,000	pa
Additional rent	£80,000	pa
YP 8 yrs @ 3.5%	6.8740	
	£549,920	

Following 7 years			
Rent payable	£200,000 pa		
YP 15 yrs @ 3.5%	11.5174		
Less YP 8 yrs @ 3.5%	6.8740	4.6434	£928,680
NPV			£1,478,600
Less exempt amount			£150,000
SDLT@ 1% on			£1,328,600
SDLT payable			£13,286

G. Surrender and re-grant, and sale and leaseback

i. Surrender and re-grant

6.30 Where a lease is surrendered in return for a new lease between the same parties, SDLT is payable on the rent reserved in the new lease. Any value in the lease surrendered is ignored (Schedule 17A of the FA 2003, para 16). However, any rent that was payable under the surrendered lease is deducted from the rent reserved under the new lease in determining the NPV, similar to a variation of a lease in respect of rent (Schedule 17A, para 9 and see Example 6.7).

Example 6.9 Surrender and re-grant of a lease

As for Example 6.8, but lessee surrenders existing lease and takes a new lease for 15 years at a rent of £200,000 pa.

Calculate NPV
The same as in Example 6.8.

ii. Sale and leaseback

6.31 The sale or an assignment of an interest, whereby the vendor or assignor takes a lease of the premises the subject of the deal, is not uncommon, particularly where the owner wishes to raise capital to use in the business. For example, a multiple retailer will sell the freehold interest in a store, where the yield is, say, 5%, to invest the proceeds in the business, where the return might be 15%.

6.32 In these cases, the purchaser or assignee will pay SDLT. However, no SDLT will be payable in respect of the new lease, subject to certain conditions. These are, that the premises which are sold are the same as the demised premises in the lease, the consideration is in the form of cash or the release or assumption of a liability, and that the parties to the sale and the lease are the same. This does not apply if the parties are companies within a group (Schedule 17A of FA 2003).

H. Amount of tax payable — residential

i. Property not within a designated disadvantaged area

a. Consideration is only rent

6.33 SDLT is payable by the lessee on the consideration paid for the lease. The provisions regarding calculation of tax payable in respect of leases are contained in Schedule 5 to FA 2003. Where the consideration is rent only, the tax is assessed on the NPV, "the relevant rental value". The tax payable, if the property is entirely residential, is 1% of the NPV in excess of £60,000, or £125,000 after 16 March 2005, (Schedule 5, para 2(3)).

6.34 If the lease is one of a number of linked transactions, the aggregate NPV of each lease is calculated, including the NPV of the lease in question, the TNPV. The tax payable as if this were a single transaction is then calculated. That amount is multiplied by:

$$\frac{\text{NPV}}{\text{TNPV}}$$

so that there is a single exempt band of £60,000 or £125,000 for all of the leases (Schedule 5, para 2(6)).

b. Consideration is rent and other payments

6.35 Where a lessee takes a lease whereby, in addition to rent, some other payment is made, such as a premium, the SDLT payable on the rent element remains 1% of the NPV in excess of £60,000 or £125,000. The SDLT on the premium is as described in Chapter 4, with the rate of the tax being 0, 1, 3 or 4%, depending on the amount. However, if the annual rent payable exceeds £1,000 per annum, the 0% band is lost, and payment up to £60,000 or £125,000 comes within the 1% band (Schedule 5 to FA 2003, para 9).

6.36 The annual rent means the average annual rent over the term of the lease. If the lease transaction is one of a number of linked transactions, the annual rent is the total of the annual rents payable under all of the transactions. Where the rent varies for different periods of the term, and the different rents are ascertainable at the start of the lease, the annual rent is the average annual rent over the period when the highest rent will be payable.

ii. Property within a DDA

a. Consideration is only rent

6.37 If the residential property is within a designated disadvantaged area (DDA), and only rent is payable, the nil band is up to £150,000 in place of up to £60,000 or £150,000 in other cases, and SDLT is payable on the NPV in excess of £150,000.

b. Consideration is rent and other payments

6.38 If the consideration is partly rent and some other consideration, such as a premium, if the NPV of the rent does not exceed £150,000, no tax is payable on the rent consideration.

6.39 Similarly, if the rent does not exceed £1,000 per annum, and the other consideration does not exceed £150,000, no tax is payable on the other consideration. (Finance Act 2008 (FA 2008))

6.40 Where the rent does exceed £1,000 per annum, there is no 0% band, the same as for property outside the DDA, so that other considerations up to £150,000 attract tax at 1% (Schedule 6 to FA 2003, para 5 as amended by FA 2008).

I. Amount of tax payable — non residential

6.41 For the purposes of determining SDLT payable on non-residential property, such properties include those where there are mixed uses, including some residential use (but see para 6.43 below).

i. Property not within a DDA

a. Consideration is only rent

6.42 As for residential property, the tax is assessed on the NPV, the amount payable being 1% of NPV in excess of £150,000 (Schedule 17A to FA 2003, para 3). The provisions regarding linked transactions referred to in para 6.34 apply equally to non-residential property.

b. Consideration is rent and other payments

6.43 As for residential property, SDLT on the rent element is separate from the tax on the other consideration. Also, as in that case, if the rent payable under the lease exceeds £600 per annum, the 0% band is lost and payments up to £150,000 come within the 1% band.

ii. Property within a DDA

6.44 If none of the property is residential, then no SDLT was payable on rent or other consideration on transactions before 17 March 2005. However, if part of the property is residential, then any rent or other consideration is to be apportioned on a just and reasonable basis to determine the rent attributable to the residential part. In most cases, the parties to the lease will have agreed the rent payable for the various areas of use, so that the rent can readily be apportioned to reflect the rent for the residential part and for the other part or parts. Apportionment by value seems to be the most just and reasonable basis.

6.45 For example, if a lessee has taken a lease of a shop with a flat over at a total rent of £40,000 per annum, the rent will have been determined by considering the rent payable for comparable shops alone, say £32,000 per annum, the balance of £8,000 per annum being arrived at by considering the rent for similar flats. What is certain is that there will be few cases where apportionment on an area basis could be considered as just and reasonable. That would arise where the rental values on an area basis are at the same level for both residential and other uses, but then the value approach would give the same answer.

6.46 Having made the apportionment, the rent or other consideration for the residential part is subject to SDLT in the normal manner for residential property, as described in paras 6.37 to 6.40 above.

Tax Payable or Repayable During a Tenancy

7

A. Within first five years

i. Resolution of rent review

7.1 As explained in paras 6.15 and 6.16, a rent review within the first five years of the term is treated as a contingent event, which requires an estimate of the likely level of rent which will be agreed on review to be adopted when calculating the net present value (NPV). Example 6.1 shows how this operates.

7.2 When the rent review is settled, the original NPV needs to be recalculated if the rent is different from the predicted rent. This will lead to a consequential further payment or a repayment of stamp duty land tax (SDLT).

Example 7.1 Adjustment of tax liability following settlement of rent review

Adopt the facts of Example 6.1. A property is let before 17 March 2005 for 12 years at a rent of £10,000 per annum (pa), subject to upward only rent reviews after four and eight years.

The predicted rent after four years was £11,255 pa, producing a NPV of £104,151.

a. Assume actual rent fixed at the first review is £12,500 pa

Calculate NPV		
Initial rent	£10,000 pa	
YP 4 yrs @ 3.5%	3.6731	£36,731
Rent after 4 years	£12,500 pa	

YP 12 yrs @ 3.5%	9.6633		
Less YP 4 yrs @ 3.5%	3.6731	5.9902	£74,877
NPV			£111,608

NPV (using actual rent)	£111,608
NPV (using predicted rent)	£104,151
Increase in NPV	£7,457

Additional SDLT will be payable depending on whether the property is residential or not and is within a designated disadvantaged area (DDA) or not (adopting statutory position when lease was granted).

Note: The rent after the fifth year is taken to be the highest rent over any 12-month period in the first five years (in this case the highest rent is during year five) — see para 7.11 below.

b. Assume actual rent fixed at £10,500 pa

Calculate NPV			
Initial rent		£10,000 pa	
YP 4 yrs @ 3.5%		3.6731	£36,731
Rent after 4 yrs		£10,500 pa	
YP 12 yrs @ 3.5%	9.6633		
Less YP 4 yrs @ 3.5%	3.6731	5.9902	£62,897
NPV			£99,628

NPV (using predicted rent)	£104,151
NPV (using actual rent)	£99,628
Reduction in NPV	£4,523

This will result in a refund of SDLT if any was paid at the start of the lease.

7.3 In practice, when a valuer is acting for a lessee in respect of a rent review, the valuer should be made aware of the predicted rent adopted at the start of the lease so as to appreciate the SDLT consequences of agreeing a rent at a different level.

ii. Resolution of contingent event

7.4 Where the NPV has been calculated in relation to a future contingent event within the first five years of the lease, the NPV will be recalculated once the actual outcome is known, in the same manner as the resolution of a rent review.

B. End of fifth year of term

i. Turnover rents

7.5 If the rent payable under a lease is a turnover rent (rent being a percentage of the lessee's turnover), or a rent which is part fixed and part turnover, clearly the amount of rent that will be payable under the lease will be unknown at the commencement of the lease, when the NPV is required.

7.6 In order to calculate the NPV, an estimate of the likely level of rent that will be achieved is made, and these predicted rents are included in the calculation.

7.7 If the lease is for a term exceeding five years then, at the end of the fifth year, the highest rent paid in any continuous period of 12 months in those first five years is identified, and this is adopted as the rent payable for the balance of the term (Schedule 17A, para 7 of the FA 2003). A revised calculation is then made of the NPV with consequential further payments or repayments being made. The revised calculation adopts the rents actually paid in the first five years, and the highest 12-month figure for the remaining period.

Example 7.2 Lease with turnover rents

1. Start of lease

A shop not in a DDA is let in September 2004 for 15 years at a fixed rent of £10,000 pa and an additional payment of 5% of the turnover of business conducted from the shop.

The lessee's business plan is based on an anticipated annual turnover of £1,000,000 after the first start-up year, when the turnover will be less. If successful, turnover should increase at around 10% each year once the business is established.

It would be unrealistic to project these assumptions into annual payments for each of the 15 years of the term. The rent, based on the target turnover of £1,000,000, will be £50,000 (5%) plus £10,000 = £60,000. This rent, or perhaps something higher, to reflect optimistic outcomes, would seem to be reasonable to adopt.

Hence,	
Predicted rent, say,	£60,000 pa
YP 15 years @ 3.5%	11.5174
NPV	£691,044
Less exempt NPV Band	£150,000
Taxable NPV	£541,044
SDLT @ 1%	£5,410

2. End of the fifth year

At the end of the fifth year, the business records show that, in the 12-month period starting in June in the fourth year up to the end of May in the fifth year, turnover was £1,400,000. This produces a rent of £70,000 + £10,000 = £80,000 pa, the highest rent in any 12-month period. (In practice, identifying the highest rent over any 12-month period would depend on how frequently the turnover figures are produced.)

The rent actually paid in the first five years was:

Year 1	£25,000
Year 2	£53,000
Year 3	£60,000
Year 4	£60,000
Year 5	£74,000

Hence, recalculate NPV =			
Year 1		£25,000	
YP 1 yr @ 3.5%		0.9662	£24,155
Year 2		£53,000	
YP 2 yrs @ 3.5%	1.8997		
Less YP 1 yr @ 3.5%	0.9662	0.9335	£49,475
Years 3 & 4		£60,000 pa	
YP 4 yrs @ 3.5%	3.6731		
Less YP 2 yrs @ 3.5%	1.8997	1.7734	£106,404
Year 5		£74,000	
YP 5 yrs @ 3.5%	4.5151		
Less YP 4 yrs @ 3.5%	3.6731	0.842	£62,308
Years 6 to 15		£80,000 pa	
YP 15 yrs @ 3.5%	11.5174		
Less YP 5 yrs @ 3.5%	4.5151	7.0023	£560,184

Example 7.4 Assignment by an exempt assignor

A charity takes a lease of a shop to use as a reception for its charitable purposes. The lease is for 15 years at an initial rent of £60,000 pa with a rent review in year five. The rent is reviewed to £90,000 pa. The charity assigns the lease after eight years for £200,000 to a multiple retailer. The property is not in a DDA.

It is assumed that the assignee has taken a new lease for seven years at an initial rent of £90,000 pa.

SDLT payable		
Consideration	£200,000	
SDLT @ 1%	0.01	£2,000
Rents	£90,000 pa	
YP 7 yrs @ 3.5%	6.1145	
NPV	£550,305	
Less exempt amount	£150,000	
Taxable NPV	£400,305	
SDLT @ 1%		£4,003
SDLT payable		£6,003

Note: Strictly, the rent payable at the next review should be predicted as it is within the first five years of the notional new lease (see paras 6.15 and 6.16). As the intention is to illustrate the effects of assignments by exempt assignors, assume that no change in rent is predicted.

NPV	£802,526
Less original NPV	£691,044
Additional NPV	£111,482
Additional SDLT @ 1%	£1,114

ii. Variable rents

7.8 It is rare, nowadays, for a rent under a lease to be a fixed rent unless it is a short lease, say up to a term of five years.

7.9 Rents are varied in several ways. One is by rent review, commonly upward only to the prevailing market rental value. A second is by pre-determined increases set out in the lease. A third is variation based on changes in the retail price index.

7.10 In the case of rent review and fixed increases, any changes that will take place after the fifth year are ignored. Instead, the highest rent paid in any continuous 12-month period is adopted as the rent payable for the balance of the term after five years.

7.11 As was shown in Example 7.1, where there was a rent review after four years, this produced a new rent which was the highest rent in the first five years, and was therefore adopted as the rent for the remainder of the term.

7.12 Where a lease is granted with five yearly reviews, the initial rent, or the first annual rent paid after any rent free period, is taken as the rent payable throughout the term. An extension to this is where there is provision in the lease for a rent review, expressed as following five years after a date which falls within the three months before the lease term commences, usually the preceding quarter day, so producing a rent review in the final quarter of the fifth year. In such circumstances, the first five years of the term are taken to end on the rent review date (Schedule 17A to FA 2003, para 7A, and see para 6.16).

7.13 Where the variation is through changes in the retail price index, these changes are ignored (Schedule 17A, para 7(5)).

C. After five years

i. Abnormal increases in rent

7.14 As has been stated, changes to rents after the fifth year through rent review or stepped rent provisions or increased turnover are left out of account in calculating NPV.

7.15 However, there is provision for rent increases to be brought into the SDLT net if the increase in rent is abnormal. Her Majesty's Revenue and Customs (HMRC) introduced this provision as an anti-avoidance measure to prevent lessees exploiting the ignoring of rent changes after the fifth year.

7.16 The provision is that, if the rent increases annually by more than 5% plus Retail Prices Index (RPI), this is treated as an abnormal increase. In such circumstances, it is assumed that a new lease has been granted at a rent equal to the rent increase.

7.17 It is hard to see what avoidance has been stopped, and since the lessee has no control over the rental growth or the RPI, if a rent review occurs at a time when rental values are at a high point, then the lessee is a victim of market changes. Note, also, that it is not the rent above 5% plus RPI which is taxed, but the whole increase.

7.18 Schedule 17A of FA 2003, paras 14 and 15, set out the conditions under which a rental increase can be considered abnormal. Basically, the date of review in relation to these provisions falls at or after the end of the fifth year of the lease. These have been subject to significant change introduced by s 164 and para 8(1) to Schedule 25 of Finance Act 2006 (FA 2006), which reduced the six steps envisaged in the original legislation to three.

7.19 The question the taxpayer must ask at the point of review is whether or not the abnormal increase provisions are in point. These provisions will not affect rent reviews prior to December 2008. The legislation originally provided that, where the annual rent payable after year five increased by more than 5% plus RPI, over and above the rent used in the original Land Transaction Return (LTR), then such increases were regarded as abnormal and would be treated (deemed) as the grant of a new lease. It can be seen that if the RPI increased at, say, 3% per annum, then an increase of 50% over, say, five years would trigger the provisions.

7.20 The provisions of para 15 to Schedule 17A of FA 2003 modified by the introduction of a formula which re application of three steps.

- Find the start date.
- Find the number of whole years between the and the date on which the new rent first become
- Test the new rent against the formula $\frac{R \times Y}{5}$

  where R is the rent previously taxed and wher number of whole years.

The excess rent must be greater than the product of the this is a slightly complex way of stating that if the rent by more than 100% then it will be treated as abnormal.

7.21 Assume a lease for 21 years is entered into on 1 January start date), subject to reviews at three yearly interval first review in January 2007 no additional (abnormal LTR is required. However, in January 2009 the rent pa agreed under the review as at 2007, must be tested to se be considered abnormal. Then, following each review 2013, 2016, 2019 and 2022, the rent must be retested to se reference to the original LTR submitted in January 2004, last occasion when there was an abnormal increase, increases can be considered and a fresh LTR may be with additional SDLT.

ii. Acquisition of lease from SDLT exempt as

7.22 In Chapter 1, various bodies were identified as being from the liability to pay SDLT. Hence, if such an exem takes a lease, no SDLT is payable by that lessee.

7.23 If the exempt body later assigns the lease, the assi acquiring a lease where no SDLT has been paid on th including the future rents that the assignee will pay.

7.24 If the assignee is not, itself, an exempt body, then SDLT payable on the NPV of the remaining rents, in additior SDLT payable on any capital payment for the lea assignment is treated as if it were the grant of a new lease unexpired term of the actual lease and on the same (Schedule 17A to FA 2003, para 11).

Interaction with VAT

8

A. Introduction

8.1 The interaction between stamp duty land tax (SDLT) and value added tax (VAT) differs from that previously in place between stamp duty and VAT. The interaction can also lead to errors being made.

B. Consideration and VAT

8.2 Chargeable consideration for the purposes of SDLT is defined in Schedule 4 to Finance Act 2003 (FA 2003). The issue of VAT is addressed at para 2 of Schedule 4:

> The chargeable consideration for a transaction shall be taken to include any value added tax chargeable in respect of the transaction, other than value added tax chargeable by virtue of an election under paragraph 2 of Schedule 10 to the Value Added Tax Act 1994 (c.23) made after the effective date of the transaction.

8.3 Therefore, unlike stamp duty, VAT only attracts SDLT, as part of the consideration, when it is actually charged. Set out below are some examples of the impact of VAT both for purchases and for the taking of leases.

8.4 Also set out below is the issue of when a transaction qualifies as a transfer of a going concern (TOGC) for the purposes of VAT. This is because, where a property subject to an election to waive exemption is sold, and the transaction constitutes a

TOGC, the transaction falls outside the scope of VAT and, therefore, the consideration is taken as being the amount actually paid net of VAT.

Example 8.1(a)

Where the purchase price of a commercial property is £1,000,000 exclusive of VAT, SDLT at 4% is £40,000.

Example 8.1(b)

Where the purchase price of the same property is subject to VAT, the gross consideration would be £1,175,000 and would attract SDLT at 4%, giving a charge to tax of £47,000.

Example 8.1(c)

A 15-year commercial lease consisting of 15 payments of £24,000 per annum (pa). VAT may be payable in addition.

Annual rent (No VAT)	£24,000	pa
YP 15 years @ 3.5%	11.5174	
	£276,417	
Less exempt band	£150,000	
Net Present Value (NPV)	£126,417	

SDLT at 1% is therefore £1,264

Example 8.1(d)

Annual rent (with VAT)	£28,200	pa
YP 15 years @ 3.5%	11.5174	
	£324,790	
	£150,000	
NPV	£ 174,790	

SDLT at 1% is therefore £1,747. (Note, due to gearing effect the amount of SDLT increases by some 38.21%.)

Example 8.1(e)

Mathematically, the gearing effect reduces as the size of the transaction increases. Therefore, if one assumes, on the basis of the facts above that the annual rent was £240,000 (exclusive of VAT) pa, the qualifying NPV in excess of the band of £150,000 would attract SDLT, at 1%, of £26,141.

Annual rent (No VAT)	£240,000 pa
YP 15 years @ 3.5%	11.5174
	£2,764,176
Less exempt band	£150.000
NPV	£2,614,176

Example 8.1(f)

However, if, on the basis of this annual rent attracting VAT the inclusive rent would rise to £282,000, the SDLT, at 1% payable on the NPV, in excess of the £150,000 exempt band, would be £30,979.

Annual rent (With VAT)	£282,000 pa
YP 15 years @ 3.5%	11.5174
	£3,247,906
Less exempt band	£150.000
NPV	£3,097,906

This equates to a 15.6% increase.

C. SDLT mitigation

8.5 It is clearly in the interests of a purchaser that VAT should not be chargeable in addition to the purchase price of the property where at all possible.

8.6 Vendors of commercial property, where the property is not subject to an election to waive exemption, or VAT incurred on the property is being recovered in the course or furtherance of the vendor's business, and, in either case, forms part of their capital goods scheme position, tend to make an election to waive exemption prior to sale. This is to ensure that all VAT incurred on the sale is properly attributable and recoverable.

8.7 Whether it is worth the vendor and purchaser striking a bargain between them with regard to the unrecoverable VAT depends on a number of factors, including their personal taxation circumstances. It also depends on the overall scale of the transaction.

Example 8.2(a)

On a transaction of £2,000,000 with fees of 2%, the VAT element in addition to the fees is £7,000. The property is not subject to an option to tax.

Purchase price £2,000,000 (no VAT)
Fees at 2% = £40,000
VAT on fees at 17.5% = £7,000
SDLT at 4% on £2 million = £80,000

Example 8.2(b)
However, if the same property were to be sold for £2,000,000 plus VAT, following an option to tax, and fees were to remain at 2%, the element of VAT in addition to those fees would still be £7,000, against an increase in SDLT from the purchaser's perspective of £14,000.

Purchase price £2,350,000 (with VAT)
Fees at 2% = £40,000 (fees not due on VAT element)
VAT on fees at 17.5% = £7,000
SDLT at 4% on £2.35 million = £94,000

8.8 Where a property comprises a property rental business and the vendor has opted to tax, as many property investors do, and where the purchaser has opted to tax prior to completion, then, under the TOGC rules, the consideration may be net of VAT after all, with a corresponding reduction in SDLT for the purchaser.

D. TOGC

8.9 A TOGC occurs when the assets of a business are transferred from one person to another, and where the purchaser of the business continues to conduct the business in the same way as the vendor. If the vendor had made a valid election to waive exemption (opted to tax), the purchaser must also have put in place a valid election to waive exemption prior to the acquisition of the property (normally before completion of the purchase).

8.10 For the purposes of a TOGC, letting property is considered to be the conducting of a property rental business. It is very important that the effect of the transfer is to put the new owner in possession of a business which can be operated as such. From the perspective of SDLT mitigation, it is also important that there must not be a series of immediate consecutive transfers of the business.

8.11 For example, if A sells its assets to B, who immediately sells the assets on to C, because B has not carried on the business, the TOGC provisions do not apply to any of the transactions. This means that the sales take on their normal VAT liability (taxable or exempt). Such immediate transfers often occur in property transactions where A contracts to sell property to B, and B sub-sells the property to C, with both contracts being completed by a single transfer by A to C.

8.12 However, different rules apply in Scotland to the transfer of a property rental business. Under Scottish law, the disposition of the *dominium utile* may be seen to be direct from A to C, and the TOGC provisions may still apply. Her Majesty's Revenue and Customs (HMRC) Notice 700/9 headed *Transfer of a business as a going concern* details what is, in HMRC's opinion, a transaction that qualifies as a TOGC.

8.13 At para 7.2 of Notice 700/9, HMRC set out some examples where they consider that a business can be transferred as a going concern. In précised form these are:

a. If a freehold of a property let to a tenant is sold subject to the existing lease, a business of property rental is transferred. This may be a TOGC, and remains so even if the property is only partly let.

b. Similarly, where a leased property (subject to a sublease) is assigned with the benefit of the sub-lease, this may be a TOGC.

c. Where a building is let, with an initial rent-free period, if the building is sold during the rent-free period, it may still be considered to be a TOGC.

d. Also, where a lease has been granted in respect of a building but occupation has not commenced, a sale may be a TOGC.

e. Where a property owner has found a tenant but has not yet entered into a lease agreement, the transfer the property to a third party, with the benefit of the prospective tenancy, prior to a lease being signed, may be considered sufficient evidence of intended economic activity for the sale to be a TOGC.

f. A property developer selling a composite site (to a single buyer) which is a mixture of let and un-let, finished or unfinished properties, and the sale of the site would otherwise have been standard rated, then, subject to the purchaser electing to waive exemption for the whole site, the sale can be regarded as a TOGC.

8.14 Paragraph 7.3 of Notice 700/9 sets out some examples where HMRC considers that there *is not* a TOGC. In précised form these are:

a. If a property developer has built a building and allowed a temporary occupation until sale (without any future occupational rights), or is actively marketing it in search of a tenant, there is no property rental business being carried on.

b. If the freeholder of a property grants a lease, it will not be a TOGC — they retain an asset (the freehold) and create a new asset (a lease). Similarly, when the owner of a head lease grants a sub-lease, it is not a TOGC.

c. When selling a property where the lease is surrendered immediately before the sale, the property rental business ceases and the sale cannot be a TOGC — even if tenants under a sublease remain in occupation.

d. The sale of a property to the existing tenant is not a TOGC, because the tenant cannot carry on a property rental business in the same way.

e. Where a tenant is running a business from the premises and sells the assets of their business as a going concern, then surrenders the lease, the grant to the new owner of the business of a lease in respect of the building is not a TOGC.

8.15 Furthermore, where a property is let and the purchaser of a property rental business is a member of the same VAT group as the existing tenant, this is not a TOGC.

8.16 Where a VAT group member sells, to a third party, a property currently being rented to another group member, this is also not a TOGC.

8.17 However, where a property rental business is sold where the tenant is a member of the outgoing landlord's VAT group, and is only one of a number of tenants, this can be a TOGC.

8.18 Similarly, where a property rental business is sold and the tenant, who is a member of the purchaser's VAT group, is only one of a number of tenants, this may also be a TOGC.

E. Anti-avoidance

8.19 Transferring a property by way of a TOGC is mandatory where the correct circumstances apply. In the Budget of 2004 (at Budget Notice CE 30/04 and BB 12/04), an anti-avoidance measure was introduced affecting the operation of TOGCs. This anti-avoidance measure was produced to attack the use of what HMRC described as artificial structures, designed either to increase the amount of input tax claimable by businesses in exempt sectors, or to spread the VAT cost of purchase over a number of years. In certain circumstances, this anti-avoidance measure removes the de-supply of a TOGC in property transactions, thereby causing the vendor's supply of the freehold to be taxable. These regulations have been slightly modified and expanded in the Finance Act 2007, but this does not affect the SDLT situation.

8.20 The measure also seeks to disapply the option to tax for a partly exempt user's special purpose vehicle, so causing the VAT on its purchase to stick with it as an actual cost and increasing the SDLT liability.

8.21 The new provision requires anyone making supplies under a lease, but who did not make the initial grant, to apply the option to tax disapplication test as though they had made a new grant. The new measure involves some amendments to para 5 of the VAT (Special Provisions Act) Order 1995, and to Schedule 10 to the VAT Act 1994 by way of statutory instruments.

8.22 The upshot of these new measures is that, while formerly the transferor of a property in a TOGC had to be satisfied that the transferee had opted to tax, there is now an additional requirement for the transferee to tell the transferor whether or not the option they have made will be disapplied as a result of the changes to para 2 of Schedule 10 to the VAT Act 1994.

F. Miscellaneous

8.23 There are references to refraining from exercising the option to tax in Chapter 19 hereof concerning the "six allowable steps" not considered hallmarks of avoidance. It is not considered to be an attempt to mitigate or avoid SDLT by virtue of not exercising the option to tax.

8.24 Where chargeable consideration is based on market value as ascertained under s 118 of FA 2003 (see also Chapter 10), then VAT should not be added to such chargeable consideration as it is based on a hypothetical transaction rather than an actual one.

Partnerships 9

A. General

9.1 Partnership transactions can take on various forms. This is recognised in the legislation, and extensive provisions are contained in Schedule 15 to Finance Act 2003 (FA 2003), introduced by s 104. The legislation as originally enacted did not fulfill the Government's intentions and has been subject to significant amendment in subsequent Finance Acts. While the provisions are still complex, they do not have as wide an impact as originally proposed.

9.2 Schedule 15 is split into three parts. Part 1 defines "partnership" and contains other general provisions. Part 2 deals with ordinary partnership transactions, while Part 3 makes provision for certain transactions (to which special provisions apply).

B. Definition of partnership

9.3 A partnership is defined as being:

a. a partnership within the Partnership Act 1890
b. a limited partnership within the Limited Partnerships Act 1907
c. a limited liability partnership within the Limited Liability Partnerships Act 2000 or the Limited Liability Partnerships Act (Northern Ireland) 2002.

9.4 An organisation formed under a law outside the United Kingdom which is of a similar character to partnerships formed under the partnership statutes, will be regarded as a partnership (Schedule 15 to FA 2003, para 1).

9.5 For the purposes of stamp duty land tax (SDLT), dealings of the partnership are treated as being by or on behalf of the partners and not by the partnership (para 2). A partnership is treated as continuing after a change in the membership so long as any one person was a member before and after the change. Since the Finance Act 2007 (FA 2007), partnerships affected by the "special provisions" are property investment partnerships (PIPs), these are defined in para 14(8) to Schedule 15 of FA 2003, as being ones whose "sole or main activities are investing or dealing in chargeable interests whether or not the activity involves the carrying out of construction operations".

C. Ordinary partnership transactions

9.6 Part 2 of Schedule 15 of FA 2003 deals with ordinary partnership transactions which are purchases entered into by or on behalf of the partners, apart from transactions (by PIPs) to which Part 3 applies (see para 9.10 *et seq*).

9.7 Partners are responsible as purchasers for observing the SDLT requirements under FA 2003. The "responsible partners" are those who were partners at the effective date of a transaction or who become partners after, and they are jointly and severally liable for any tax, interest or penalty due (paras 6 and 7 of FA 2003). In the case of partners who become a responsible partner after the effective date, no amount of SDLT is recoverable from them.

9.8 However, the partners, or a majority of them, may nominate one of their number to act as a representative partner, who will be responsible for complying with the SDLT requirements. Any nomination must be notified to Her Majesty's Revenue and Customs (HMRC) before it comes into effect, or, indeed, before revocation of a nomination comes into effect. Partnerships in general are treated for SDLT purposes, as are individuals where they acquire chargeable interests. It is the movement of property assets within a partnership, or in and out of PIPs which which give rise to apparently differential treatment.

9.9 There are some differences in relation to Limited Liability Partnerships (LLP) that should be appreciated. Under the Limited Partnerships Act 1907, a LLP must have at least one general partner who manages the business and has full liability for the partnership debts, plus one or more limited partners, who cannot be involved in management but are only liable to the extent of their original capital contributions. To restrict liability the capital contributions are usually small, with working capital provided by loans. Because of the nature of LLPs, partners cannot withdraw their capital during the partnership's lifetime. To assign a partnership share, the permission of the general partner must be obtained, the obtaining of consent amounts to "arrangements" and the payment of capital to the partnership is the giving of money or moneys worth and therefore chargeable consideration under para 36 (a) to Schedule 13.

D. Transactions to which special provisions apply

9.10 It is common within a partnership for changes to be made to a partner's entitlement to a share of the profits or to a percentage share of a partnership asset. The proportion of the profits is that person's "partnership share". This same proportion will be taken to apply to the ownership of partnership property, which is an interest held by the partnership for the purposes of the partnership business (Schedule 15, para 34 of FA 2003).

9.11 Generally, the acquisition of an interest in a partnership is not a chargeable transaction, even though the partnership property includes land (Schedule 15, para 29 — and see para 9.13). A charge to stamp duty continues on the transfer of partnership interests.

9.12 The transfer of an interest in a partnership occurs where a partner transfers the whole or part of their interest as a partner to another, who may be an existing partner, or where a person becomes a partner and an existing partner reduces their interest or ceases to be a partner (Schedule 15, para 36).

9.13 Exceptions to the general rule that the acquisition of an interest in a partnership is not a chargeable transaction (para 9.11) are where there is:

a. the transfer of a chargeable interest to a partnership (Schedule 15, para 10)
b. the transfer of a chargeable interest in a partnership for which consideration is given for the transfer (Schedule 15, para 14)
c. the transfer of a partnership interest pursuant to earlier arrangements (Schedule 15, para 17).

9.14 Further, there is a chargeable transfer where there is a transfer of a chargeable interest which was, but ceases to be, a partnership interest, or where a chargeable interest is granted or created out of partnership property which is not, itself, partnership property, for example, the grant of a lease to a third party (Schedule 15, para 37).

9.15 Where there is a chargeable transaction to, in, or from a partnership, it is necessary to allow for the fact that the acquisition may be by someone who will not acquire the whole interest in the land and that the subject of the transaction is, or will become, or will cease to be a partnership asset. Extensive and complex provisions are included within Schedule 15 to deal with this situation.

9.16 Take, as an example, the transfer of a chargeable interest to an existing partnership, or the transfer in connection with the formation of a partnership (Schedule 15, para 10). The provisions apply in three cases, namely where:

a. a partner transfers a chargeable interest to a partnership
b. a person transfers a chargeable interest to a partnership in return for an interest in the partnership
c. a person transfers a chargeable interest to a person who is connected with a partner, or with a person who becomes a partner as a result of, or in connection with, the transfer.

9.17 The chargeable consideration for the transaction, if the partnership is not formed wholly of bodies corporate, is:

$$(RCP \times MV) + (RCP \times AC)$$

where:

RCP is the relevant chargeable proportion,
MV is the market value, and
AC is the actual consideration.

9.18 RCP, in relation to the market value is (100 ñ SLP)% where SLP is the sum of the lower proportions, whereas RCP in relation to the actual consideration is SLP% (so that if SLP is, say, 30, then the chargeable consideration will be 70% of market value plus 30% of the actual consideration).

9.19 SLP is determined by a five-step calculation (Schedule 15, para 12).

Step One
Identify the relevant owner or owners. A relevant owner is a person who was entitled to a proportion of the chargeable interest immediately before the transaction, and, immediately after, was a partner or connected with a partner.

Step Two
Identify the corresponding partner of each relevant owner. A corresponding partner is any person who, immediately after the transaction, was a partner and also a relevant owner or connected with a relevant owner.

Step Three
For each relevant owner, find the proportion of the chargeable interest to which they were entitled immediately before the transaction, and apportion that proportion between the other corresponding partners of the relevant owner.

Step Four
Find the lower proportion for each person who is a corresponding partner to one or more relevant owners. This is the lower of the proportion of the chargeable interest attributable to the partner, or the partner's partnership share immediately after the transaction. The partner's proportion of the chargeable interest is the proportion apportioned to them at Step Three or, if they are the corresponding partner to more than one relevant owner, the aggregate of the proportion apportioned to them at Step Three.

Step Five
Add together the lower proportions of each person determined at Step Four, which are the sum of the lower proportions, or SLP.

9.20 The purpose of these complex calculations is to exclude from charge the proportion of the value of the land which is attributable to any transferor and which is retained by themas a partner. Similar calculations are applied to other types of transactions referred to in para 9.13.

9.21 For example, there are specific provisions in Schedule 15, para 11 for circumstances where a transfer is made to a partnership, and the chargeable consideration includes rent, to separate out the rental element. Similar conditions apply in regard to transfers from a partnership (Schedule 15, para 19).

9.22 Generally, all reliefs and exemptions, including disadvantaged area relief (DAR), apply to partnership transactions (Schedule 15, paras 25 and 26), together with both group and charities reliefs (Schedule 15, paras 27 and 28).

9.23 Certain transactions are not notifiable and, again, these follow the general provisions in respect of SDLT (Schedule 15, para 30).

9.24 It should also be appreciated that stamp duty still applies to partnership transfers and modifications of interests (Schedule 15, paras 30, 31 and 32).

9.25 In the case of a partnership consisting wholly of corporate bodies, if SLP is 75% or more, the consideration is taken to be the market value of the interest transferred (Schedule 15, para 13). This affects transfers into (para 10) and out of (para 18) a partnership.

9.26 The original provisions in FA 2003 were simpler. For example, the transfer of an interest in land to a partnership in return for an interest in the partnership was originally exempt from SDLT. It was amendments introduced by Finance Act 2004 (FA 2004), amending FA 2003, which led to these complexities. It is worthy of note that the calculations require a figure derived from a valuation. History has shown that a tax which depends on the uncertainty of a valuation is, itself, unpredictable.

E. Anti-avoidance

9.27 HMRC was concerned that artificial manipulation of partnership shares could lead to avoidance. For example, a partner gifts the majority of their share for no consideration and by making their share small, reduces the consideration (based on market value) correspondingly small. After the transaction is complete the share is gifted back to the taxable partner. Paragraph 17 to Schedule 15 of FA 2003 charges the transfer of the partnership share to SDLT, irrespective of the amount of consideration, on the basis that the subsequent transfer is a "linked transaction" with the original transfer of the chargeable interest into the partnership.

10 Market Value

A. Introduction

10.1 The statutory basis of valuation for stamp duty land tax (SDLT) is significantly different from that which might be adopted in the market place to ascertain value or consideration in a transaction. Therefore, when subject to Her Majesty's Revenue and Customs (HMRC) investigation, challenge or enquiry, taxpayers may find that the approach to valuation adopted provides a different answer from the one used in the particular transaction.

10.2 In many instances this will not be the case, particularly when dealing with individual properties. However, in larger and more complex transactions, a significant level of difference may be apparent because of the approach adopted by the Valuation Office Agency (VOA) and HMRC. A point which should be remebered is that where a SDLT liability is assessed on the basis of a valuation, rather than an actual sale price, it is never appropriate to make an addition for "notional" value added tax (VAT), as the value adopted is based on a hypothetical not actual transaction.

B. Statutory definitions

10.3 For the majority of the SDLT transactions, "market value" is defined in s 118 of the Finance Act 2003 as being "determined as for the purposes of the Taxation of Chargeable Gains Act 1992 (c.12) (see ss 272 to 274 of that Act)". Market value relating to the ascertainment within a land transaction, of the chargeable consideration, which relates to the chargeable interest transferred.

10.4 This approach is somewhat unusual in itself. However, it is even stranger when one considers that s 272 is titled "Valuation: general", s 273 deals with "Unquoted shares and securities", and s 274 deals with the "Value determined for inheritance tax" purposes.

10.5 In terms of SDLT, which deals with land and property, the relevant definitions are confined to s 272(1) and (2), although s 273 has relevance to the ascertainment of the market value of shares in a single property company.

10.6 Section 272(1) states: "In this Act market value in relation to any assets means the price which those assets might reasonably be expected to fetch on a sale in the open market."

10.7 Section 272(2) states: "In estimating the market value of any assets no reduction shall be made in the estimate on account of the estimate being made on the assumption that the whole of the assets is to be placed on the market at one and the same time".

10.8 The above basis of valuation has been much modified by case law and can, in certain circumstances, provide a significantly different answer from a "normal" open market valuation. The basic approach, which is mainly adapted from estate duty and inheritance tax case law, is aimed at the maximisation of value. In some respects it flies in the face of economic reality. It is this point which may create the most difficulty in practice, where the value used in any transaction is subject to HMRC challenge.

10.9 In the market place, deals are struck for a variety of reasons. Some are wholly commercial; some are born out of necessity. The statutory basis of valuation makes no allowance for a forced sale or the acceptance of an offer which, while low, is one that can be transacted immediately.

C. Basis of valuation for SDLT purposes

10.10 A proper application of the statutory rules, and an understanding of the basis of market value for taxation purposes, is of the greatest importance in preparing a valuation for SDLT purposes. The basis is actually derived from the two primary direct taxes that require property valuations — capital gains tax (CGT) (corporation tax on chargeable gains for companies) and inheritance tax (IHT), with the CGT definition adopted for SDLT purposes.

10.11 The statutory definition of market value for CGT and IHT are both written in similar terms, but they provide only a very bare definition. The statutory definitions are similar to those used in earlier tax legislation and, over the years, case law has established that, in arriving at "market value", the following assumptions have to be made:

a. The sale is a hypothetical sale.
b. The vendor is a hypothetical, prudent and willing party to the transaction.
c. The purchaser is a hypothetical, prudent and willing party to the transaction (unless considered a special purchaser).
d. For the purposes of the hypothetical sale, the vendor will divide the property to be valued into whatever natural lots will achieve the best overall price.
e. All preliminary arrangements necessary for the sale to take place have been carried out prior to the valuation date.
f. The property is offered for sale on the open market by whichever method of sale will achieve the best price.
g. There is adequate publicity for advertisement before the sale takes place so that it is brought to the attention of all likely purchasers.
h. The valuation should reflect the bid of any special purchaser in the market (provided they are both willing and able to purchase).

D. Analysis of basis

10.12 In order to analyse the meaning of "market value" and its implications, the broad definition of market value has been broken up into its component phrases, which are then examined in the light of relevant case law.

10.13 The statutory definition, in Taxation of Chargeable Gains Act 1992 (TCGA 1992), broadly defines market value at the relevant date for taxation purposes as:

> The price which the property might reasonably be expected to fetch if sold in the open market at that time, and that price must not be assumed to be reduced on the grounds that the whole property is to be placed on the open market at one and the same time.

i. The property

10.14 In *Duke of Buccleuch* v *Inland Revenue Commissioners* [1967] 1 AC 506, it was held that the reference to the property was not a reference to the whole estate being valued; it meant any part of the estate, which it was proper to treat as a unit for valuation purposes. Similarly, in *Earl Ellesmere* v *Inland Revenue Commissioners* [1918] 2 KB 735, it was held that the market price was a price based on the separate values of the various parts. It was also indicated that the price must be estimated on the basis that the properties were sold in whatever lot (or lots) would realise the best price.

10.15 In *Inland Revenue Commissioner* v *Gray (Executor of Lady Fox dec'd)* [1994] STC 360, it was held that the property must be valued as it actually existed even if, in real life, a vendor would have been likely to have made some changes or improvements before putting it on the market. Although this case referred to variations in the way in which the property was held by the parties (rather than physical works), it identified the general principle of valuing the property as it stands at the valuation date.

10.16 As a consequence, in preparing any SDLT valuation, it is important to have proper regard to the most viable lotting of the property (or properties) to be valued in order to maximise the overall price. This is, effectively, a notional marketing exercise, commonly referred to as prudent lotting.

ii. The price

10.17 In *Buccleuch*, the price which the property might have reasonably expected to fetch was defined as a gross sale price for the property without deducting any selling costs.

10.18 Furthermore, in *Ellesmere*, the price was held to mean the best possible price that would be obtainable in the open market if the property was sold in such a manner (and subject to such condition) as might reasonably be calculated to obtain for the vendor the best price for the property.

10.19 However, it is not to be assumed the best price is automatically the highest possible price that could be achieved. What is required, in valuation terms, is an estimate of the price which could be realised under the reasonable competitions of an open market on a particular date.

10.20 When carrying out an investment valuation it is common practice to arrive at a "gross" amount by capitalisation of income streams, then to make a deduction for SDLT and other purchaser's costs to arrive at a "net" value, which is the market value or amount to be paid. This is a valuation methodology seeking to arrive at market value; it is not, of itself, precluded by the above concept.

iii. If sold

10.21 The statutory definitions of market value are concerned with a hypothetical sale, not an actual one. As originally held in *Re Crossman* [1937] AC 26 (and confirmed unanimously in *Buccleuch* and also in *Lynall* v *Inland Revenue Commissioners* [1972] AC 680), it is irrelevant, in arriving at the sale, to consider what would have been the circumstances attending an actual sale.

10.22 What the property would have actually realised in the open market, or the potential and possibility of putting property on the market at the valuation date, are also irrelevant. In other words, one does not have to assume the property had actually to be sold, as a hypothetical market must be assumed as at the valuation date.

10.23 In *Gray* it was said that the property must be assumed to have been capable for sale in the open market, even if in fact it was inherently unassignable or held subject to restrictions on sales. The relevant question is what a purchaser would have paid to enjoy whatever rights were attached to the property at the relevant date, assuming the hypothetical sale.

iv. In the open market

10.24 In reference to *Lynall*, it was held that the property must be valued on the basis of a hypothetical sale between a hypothetical willing vendor (not the actual owner of the property in question) and a hypothetical willing purchaser. The hypothesis used was that potentially no one was excluded from buying (the definition of the hypothetical purchaser thus potentially including even the actual owner).

10.25 The statutory definition referred to "the open market" and not "an open market". This has been interpreted as meaning a real market made up of real people. In *Lynall*, the open market was regarded as a blend of reality and hypothesis. It was held that the conditions under which the hypothetical sale is deemed to take place should be built on a foundation of reality, so far as is possible. However, it was deemed even more important not to defeat the intentions of statute by an undue concern for reality, in what is essentially a hypothetical situation.

10.26 Case law has added additional refinements to the components of the open market and, in particular, to those parties to be assumed to be active in it. Again, in the *Lynall* case, it was held, and statute implied, that there had been adequate publicity or advertisement before the sale, and that steps had been taken (before the sale) to enable a variety of persons, institutions or financial groups to consider what offers they would be prepared to make.

10.27 However, in *Gray*, it was said it could not be emphasised too strongly that, although the sale is hypothetical, there is nothing hypothetical about the open market in which it is supposed to have taken place. The hypothetical sale envisaged (in order to ascertain the open market value for taxation purposes) presupposes a willing vendor and a willing purchaser.

v. Willing vendor

10.28 A willing seller is one who is prepared to sell provided a fair price is obtained. It does not mean a vendor is prepared to sell at any price and on any terms; in short, the hypothetical vendor is assumed to be both a reasonable and a prudent person.

vi. Willing purchaser

10.29 The principle is that the open market includes everyone who has the will and the money to buy. It has been said that the buyer, like the vendor above, must be a person of reasonable prudence and also hypothetical.

vii. Special purchaser

10.30 The case of *Inland Revenue Commissioners* v *Clay* [1914] 3 KB 466 effectively established that where there is a known purchaser in the market who is willing to buy at a considerably higher price than anyone else, (thereby enabling a vendor practically to rely on extorting that price from them), then the value of the asset for tax purposes is represented by the higher price that the special purchaser is willing to pay, or by a close approximation thereto.

10.31 In *Walton's Executor* v *IRC* [1996] STC 68, CA, it was held that it was a question of fact — to be decided by evidence — whether or not there were any special purchasers in the market, and what price they would be prepared to pay.

10.32 The concept of the reflecting of the bid of a special purchaser is a difficult one for valuers, used to ignoring the existence of a purchaser with a special interest, in preparing conventional valuations. While there is a subtle difference between a special purchaser and a purchaser with a special interest, the concept can potentially cause difficulty in an SDLT scenario, particularly where the transaction is between connected persons.

10.33 This is because the value of the assets may be greater in the hands of the hypothetical purchaser than in the market as a whole, particularly where, for example, in the case of a site assembly, that value relates to the asset's marriage value with other land. In that respect, the actual purchaser who owns the other land may potentially be considered a special purchaser.

viii. At that time

10.34 As this basis of valuation is generally applicable for taxation purposes, the time is defined by each statute for the purposes of the valuation exercise in question. The assumption regarding the definition of the date is that one has to envisage a hypothetical sale, in which all the preliminary arrangements had been made prior to the valuation date, so that the sale can take place at the statutory point in time.

10.35 The objective is to ascertain the value of the asset at the prescribed time (and not at any other time), and this can only be achieved by assuming that all preliminary arrangements had been made beforehand.

10.36 For SDLT purposes, there are two points in time in which a valuation might be made. The first is on completion of the transaction; the second is the point at which the contract is substantially performed. This is referred to in the legislation as the "effective date".

10.37 The initial definition of "effective date" appears in s 119 of FA 2003, which is "the date of completion" except as otherwise provided. The "otherwise provided" refers to the provisions of ss 44(4) and 46(3) of FA 2003 which deal with substantial performance of a contract before completion and conveyance of contracts to third parties, and options and rights of pre-emption respectively (see Chapter 1).

10.38 This section is further amended by para 8 of Schedule 39 to Finance Act 2004 (FA 2004) to include references to agreement for leases (including missives of let not constituting a lease in Scotland), followed by substantial performance, and the assignment of an agreement to lease after substantial performance, the detailed provision being contained in Schedule 17A to FA 2003, paras 12A, 12B and 19.

10.39 The definition of substantial performance is contained in s 44 of FA 2003 and includes both the taking of actual possession and the right to receive rents and profits. There are also provisions which include connected persons within the scope of the definitions. The primary element of substantial performance, however, is the payment of a substantial part of the consideration, a substantial part being considered by HMRC to be 90% or more.

10.40 The above valuation assumptions do not apply in the case of valuing a dwelling the subject of shared ownership leases (Schedule 9 to FA 2003, para 4A(4)) which provides that "Section 118 FA 2003 (meaning of "market value") does not apply in relation to the references in this paragraph to the "market value of the dwelling".

10.41 What this means is that an interest in shared ownership property is valued by taking a percentage of the property's open market value with vacant possession. To use the basis of valuation described above could lead to significant distortion in what should be a straightforward calculation.

10.42 The valuation assumptions and approach, adopted for the purposes of taxation valuation, can make a material variation in relation to valuations carried out in an ordinary open market place. In the main, this is not applicable to single assets, but the more complex and larger the transaction, the more likely that a material variation could be achieved. It should be borne in mind that the anti-flooding provision of s 272(2) of TCGA 1992 flies in the face of economic reality, and also that, for taxation purposes generally, there is no concept of a forced sale or sale at undervalue due to immediate financial pressure.

E. Evidence of value — valuation methodology

10.43 Valuers should also note important peripheral matters; it is a matter of fact that the best primary evidence is actual transactional or market evidence, although in the case of historic valuations this poses its own problems. While evidence of value can be deduced from many sources, the VOA has a database of evidence collected from a range of confidential sources, and sometimes uses the evidence of valuations agreed or determined in other cases. In the Lands Tribunal case of *Newman (HMIT)* v *Hatt* [2002] 1 EGLR 89, the member reviewed the usability of evidence obtained from confidential sources. However, it should be borne in mind that, from 1 April 2000, any person is entitled, subject to certain conditions and on payment of a fee, to inspect and make copies of Land Registry entries, including the price paid on the last change of proprietorship (Land Registration Act 1988 SI (1) and the Land Registration (No 3) Rules 1999).

10.44 The VOA guidance manuals advise district valuers that, while settlements in comparable cases constitute secondary evidence, they can be used when there is insufficient primary evidence of open market transactions. Settlements in tax cases are usually made within the context of the taxpayer's own affairs, and are often not a true reflection of market value.

10.45 Methods of valuation are not predetermined by statute or by case law; transaction evidence may reflect the application of a variety of valuation methods, but that does not affect the comparability of the price realised in each case. In practical terms, property assets are valued or appraised by whichever method is most appropriate.

10.46 Special classes and categories of assets will be appraised in different ways, using different techniques, because of the way in which the market values them. A presumption of the *Lynall* case was that the vendor — when advertising the property — makes such information available to the purchasers of that type of asset as they would expect to receive, or to be made available, in a normal market transaction. It should be borne in mind, therefore, that in this area of valuation there is no presumption of a restricted or closed sale between connected persons.

F. Apportionments

10.47 In certain circumstances apportionments are required, particularly in relation to "composite transactions" where the consideration relates in part to a land transcation and in part to other matters and, secondly, in relation to the mixed use of property. This might be where only part of the land being transferred is residential property (as defined by s 116 of FA 2003) and, where disadvantaged area relief (DAR) is in point, the consideration must be apportioned on a just and reasonable basis between the residential and non-residential elements.

10.48 The primary circumstance where apportionments are required derives from para 4 of Schedule 4 to the FA 2003. This is headed "Just and Reasonable Apportionment" and relates to circumstances where the consideration relates to two or more land transactions, in part to a land transaction and in part to another matter or to, in part, chargeable consideration and in part to other matters. Therefore, in a transaction, such as the sale of a hotel, where the consideration may relate to the "real property" asset, the chattels and to possibly "goodwill". It is the purpose of the legislation that the individual components of the transaction will be aggregated and then re-apportioned to their component parts, some of which may be taxable to SDLT and some of which may not. This statutory provision overides the intention of the parties, even in an arms length transaction where the quantum of value of the individual components may have been separately agreed. The current HMRC approach in this area, particularly in relation to "goodwill", has caused significant problems in practice and is dealt with in detail below.

10.49 "The 'just and reasonable' test is necessarily subjective, and each case will be considered on its merits", to quote the (then) Inland Revenue Statement of Practice SP 1/4, which deals with mixed use properties. It is suggested therein that an apportionment might be on the basis of the percentage areas quoted in planning applications, where appropriate, or alternatively of floor space relating to the respective uses. Other methods of apportionment will be considered as part of a claim.

10.50 There is a general requirement that apportionment for taxation purposes should be on a just and reasonable basis. It should be remembered that this has equal application both to the taxpayer and HMRC.

10.51 There is no particular just and reasonable method of apportionment laid down in statute. However, the objective is to arrive at the contribution which each part makes to the sum to be apportioned, whether that sum is a reflection of open market value or actual sale consideration in an arms length transaction.

10.52 Apportionment by area will only be appropriate where a value is spread evenly throughout the land, as was stated *obiter* in *Salts* v *Battersby* [1910] 2 KB 155. This case dealt with a method of apportioning rent, and, in that circumstance, Darling J stated that the correct approach was to be value, and not area based:

> It seems to me clear from the authorities that what you have to regard is not the bare acreage of the severed portions of the land demised, but their relative values. You find the same principle running through them all. The County Court Judge was of the opinion that the proper way to apportion rent was to have regard to the yardage and the yardage alone. Therein I think he was wrong. If it can be shown that the land is of equal value throughout, no doubt the apportionment must be on the basis of yardage. But yardage cannot be a sufficient test of the relative value by itself; and here, so far from there being evidence there was an equal value throughout, the evidence was the other way.

10.53 A more recent case involving an apportionment of part of a purchase price of an area of land was the case of *Bostock* v *Totham* (1997) 69 TC 356; [1997] STC 764. In this case, in upholding the Special Commissioners' decision, the Court suggested that the approach adopted did no more than apply the principles of the value based rateable apportionment where there is a part disposal, and was appropriate to the circumstances of the case.

G. UK Guidance Note 3

10.54 Guidance Note 3 (UK GN 3) of the Royal Institution of Chartered Surveyors (RICS) *Appraisal and Valuation Standards*, 5th ed, "The Red Book", contains the basis of valuation to be adopted in taxation purposes as set out above. In general, HMRC expects to receive properly worked through valuations, carried out on the correct basis, by competent valuers, in much the same way as other valuations carried out in accordance with the directions and recommendations of the manual. It must be remembered that this is the only GN with application to UK tax valuations.

H. Treatment of "goodwill"

10.55 The Finance Act 2002 (FA 2002) introduced a new code for the taxation of intellectual property, goodwill and other intangibles for property. Paragraph 4(2) of Schedule 29 to FA 2002 provides that goodwill has the meaning it has for accounting purposes. In the context of FA 2002, goodwill is also treated as an intangible fixed asset. Financial Reporting Standard 10 (FRS 10) used the term "purchased goodwill", which was the difference between the cost of an acquired entity and the aggregate of the fair value of the entities' identifiable assets and liabilities. International Financial Reporting Standard (IFRS) 3 considers that the consideration paid for a business may be allocated to any or all of the following components.

- Land and buildings.
- Fixtures and fittings.
- Inventory, if not separately identified in the terms of the transfer.
- Intangible assets and liabilities.
- The value of licences, etc.

10.56 In the HMRC *Capital Gains Tax Manual* (CGTM) a number of different types of goodwill were, historically, identified namely:

a. personal: relating to the skill and the personality of the proprietor of the business

b. inherent: relating to the location of the businees premise

c. free: relating to the overall worth of the business subdivided into:

    (i) free adherent goodwill: this arises not from the location of the premises but from the carrying on of a particular business for which those premises have been or are specifically adapted or licensed
    (ii) free separable goodwill: true free goodwill which is entirely separate from the business premises and can be transferred independently from them.

10.57 HMRC have claimed that where a business is sold as a fully equipped operational entity, which comprises the business together with the premises, then there can only (or mainly) be inherent and free adherent goodwill present. For SDLT purposes, anything which forms part of the property asset is taxable, any intangible items which, in this case, would include free separable goodwill, are not taxable to SDLT. Therefore, where HMRC can claim all, or the majority of the goodwill is either inherent and, or, adherent to the property, the amount taxable to SDLT as chargeable consideration increases and the tax benefits bestowed by the FA 2002 in relation to goodwill and its treatment under the intangibles regime is reduced in its effect.

10.58 HMRC set out, at CG68456 (CGTM), a list of businesses which are likely to have inherent goodwill these are:

- hotels
- public houses
- restaurants
- residential care and nursing homes
- petrol filling stations and garages retailing petrol
- mineral undertakings including mines, quarries, etc
- land infill sites
- leisure industry premises including: cinemas, nightclubs, casinos, bingo halls, theatres, bowling alleys, sports ground, theme parks
- caravan parks.

Many of these categories of property are commonly valued as "trade related properties". Adopting the approach set out in GN 1 of the "Red Book", initially HMRC took great comfort from the wording of GN 1, particulary as it used (prior to 2007) terms such as inherent goodwill, personal goodwill and the like. HMRC formed the view that where properties where valued on this basis, all (or most of any) goodwill must be inherent or adherent and therefore incapable of separation from the property asset.

10.59 HMRC pursued a case before the Special Commissioners, *Balloon Promotions Ltd* v *Wilson (Inspector of Taxes)* [2006] STC (SCD) 167, read in March 2006. This related to the treatment of goodwill for CGT purposes where only personal goodwill and free separable goodwill are recognised (by HMRC) for the purposes of the TCGA 1992. The Special Commissioner highlighted the different interpretations of goodwill between HMRC and the RICS and found that goodwill should be looked at as a whole and its exact composition would vary from business to business. While goodwill can only exist in the presence of a business, nevertheless, the Commissioner said, it can be sold separately from the premises in which the business is carried on.

10.60 This undermined the HMRC stance on goodwill and, for the purposes of SDLT, allowed taxpayers to argue that HMRC's approach was flawed and that the "total" consideration could be apportioned to types of goodwill (personal, free separable goodwill, and other intangibles) which would not be taxable to SDLT as intangible assets.

10.61 HMRC, needless to say, does not care for this view, and has continued to seek to argue that in many circumstances there can be virtually no element of free goodwill present. Therefore, almost all cases where a chargeable consideration is made up of a "composite" value are subject to investigation and enquiry. Following the Commissioner's comments in the Balloon case, from the RICS perspective it became apparent that GN 1 was being interpreted as providing guidance in a way it was not intended to do. It is considered by the RICS that there is no property valuation concept of free goodwill. GN 1 was therefore re-written to deliver those elements of guidance required for the valuation of trade related property and to include the allocation principles of IFRS 3. GN 1 makes certain assumptions which are possibly not permitted in the context of SDLT. The definition of chargeable interest in s 48(2) of FA 2003 specifically excludes as example "licences to use or occupy land". However, GN 1 assumes that all licences, permits, etc. required for the operation of the business, some of which are personal to the operations, will be transferred or granted at the date of transfer, from vendor to purchaser.

10.62 The basic premise being that because a particular category of property is valued in a particular way, it does not prove or disprove the potential existence of free separable goodwill. However, the most important element to consider is the definition in GN 1 of transferable goodwill, this is considered to be an intangible asset and is defined as such. HMRC has chosen to read the definition, which includes the word inherent as being supportive to their arguments, however, this is wholly incorrect. This point has been made to the VOA by the RICS and practitioners are advised that any VOA or HMRC arguments to the contrary are potentially flawed. In 2007, the Court of Appeal, in the case of *Condliffe* v *Sheingold* [2007] EWCA Civ 1043, held that that judge at first instance did not seem to appreciate the fact that the goodwill related to the business, and that it could be separated from the lease and be the subject of separate ownership. This case related to the residue of a three-year lease of a restaurant which went into liquidation. This apparently completely refutes the HMRC position, but doubtless they will protest otherwise.

10.63 Although HMRC is greatly concerned with "goodwill", there is no specific or statutory need to identify it for SDLT purposes. SDLT is concerned with the taxation of certain tangible assets linked to or forming part of an interest in land. Intangible assets, of which goodwill is one, are not taxable to SDLT. Therefore, so long as a differentiation can be made between the tangible and taxable assets (chattels although tangible being excluded) and "other" intangible and non-taxable assets, this should be sufficent for the purposes of SDLT. HMRC is also concerned with "goodwill" in CGT cases, however, there is no particular reason to accept an HMRC challenge to an apportionment because Shares and Assets Valuation, the HMRC group which deals with the issue of goodwill, believe the amount attributed to goodwill is mis-stated, the issue is whether the taxable and non-taxable elements are properly separated. For simplicity, the author will continue to refer to the intangible element as goodwill.

10.64 From an apportionment perspective, there is no particular requirement to value each individual element or component to use in a rateable basis of apportionment. The practice of taking any valuation differential between a property valued as a fully equipped operational entity and an empty property having regard to its trading potential as being the element of free goodwill, is technically flawed.

10.65 When seeking to identify the various component parts and quantifying their individual values for the purposes of para 4 of Schedule 4 to the FA 2003, a valuer would, probably, be best advised to adopt a *Findlay Trustees* v *CIR* CS 1938, 22 ATC 437 based approach, effectively a deduction of the tangible assets from the total consideration to identify a residue, potentially comprising or including goodwill. Or for an accountant to value the goodwill or intangible element separately, it is possible that there will be a further addition to UK GN 3 dealing with this point in due course.

10.66 There is a school of thought that suggests that when apportioning value between the taxable and non-taxable elements, one should value the property together with its chattels and have regard to the point in time that the property could be brought up to a level of fair maintainable trade. Effectively discounting the property value of a fully equipped operational entity, as achieved by adopting one of the valuation bases set out in GN 1. In the context of the valuation of a property on the statutory basis of a hypothetical transaction, this approach is flawed, the fact that one can adapt valuation assumptions to achieve a particular result does not prove that the resultant "answer" is correct. It must always be remembered that the meaning of "market value" is defined for the purposes of SDLT in s 118 to FA 2003, and that the statutory requirements are paramount. As a matter of interest, in *Condliffe* v *Sheingold*, mentioned above, the chattels were lotted with the business not the property.

10.67 There are two issues which cause further problems in practice. First, there is a general misunderstanding that a valuation carried out under GN 1 is a valuation of the property as a going concern. It is not, a GN 1 valuation is made on the assumptions set out in the Guidance Note, it relies on the concept of fair maintainable trade (FMT) this is not the actual trade being achieved by the current owner but an assessment of the trade that might be conducted by an average competent operator.

10.68 A going concern valuation relates to the actual trade being carried out by the actual operator, which might be greater than FMT or less than FMT. The problem arises that because properties normally valued by reference to their accounts are valued for loan security purposes on a GN 1 basis, often their sale and purchase price has been arrived at in this way. Therefore, the price paid is an estimate of the property's value in the hands of a notional, rather than an actual operator. It would seem to the author that this would not be a proper way to ascertain any element of "goodwill" present in the actual sale with any degree of accuracy.

I. Contingent, uncertain or unascertained consideration

10.69 The provisions of s 51 of FA 2003 are dealt with in Chapter 4. They give rise to an unusual situation in valuation terms. When valuing contingent or uncertain consideration in the open market, a valuer normally makes allowance for risk, deferment and the period over which future development might take place if the contingent event occurs. Section 51, with effect from July 2004, does not allow for risk or deferment to be taken into account. An estimate must be made as to the value as if the contingencies do not stand in the way of development.

10.70 However, the anti-flooding provisions of s 272(2) of TCGA 1992 bring another element into play. Take, as an example, a transaction comprising the sale of a large area of land, say, 100 acres. If various events occur in the future, say, the construction of a new bypass five years hence, creating a different boundary within which development may occur, and the construction of a new sewer, in about 10 years, then the whole of the land will be developable.

10.71 A transaction is entered into with an upfront payment, plus a percentage of open market value, payable at the earliest 10 years hence, if development occurs, or is able to occur. An estimate of this value falls to be made, probably based on its current development value. However, the locality will not stand the introduction of 100 acres of, say, residential development. At this point, in real terms, the land will be developed in stages over a period of three to five years, with development keeping pace with demand. Development could begin in a number of ways. In the real world, a discount over time might be made, as there is a holding cost associated with acquiring the land upfront followed by sequential development. Therefore, the gross value might be discounted by 10–20%, dependent on the current costs of finance.

10.72 The provisions of s 272(2) of TCGA 1992 make no allowance for this, and the value in this context would apparently be an estimate of the future gross market value. However, it is not readily possible to value out into the future, and it would probably be more appropriate to take the current gross market value on the assumption that the contingent events had already occurred. Similarly, given the anti-flooding provisions of s 272(2), no discount could be made for developing over time.

Example 10.1

Acquisition costs for 100 acres at, say, £5,000 per acre = £500,000.

Full development value per acre if planning consent available at today's date, say, £500,000 per acre = £50,000,000.

Future payment to be based on 80% of full development value less purchase price paid.

Therefore, future purchase price £50,000,000 x 0.8 = £40,000,000 less £500,000 = £39,500,000.

Figure to return in Land Transaction Return (LTR £39.5 million.

10.73 Valuers will appreciate that this approach is fundamentally flawed for a variety of reasons. It is, however, more appropriate than attempting to guess the extent to which values may rise or fall in the future, particularly when the contingent events on which the uplift depends may not happen within the time scales estimated or, indeed, not happen at all. An alternative might be to assume growth occurred in line with projected inflation, although this, too, is flawed.

10.74 In any event, it is probable that the SDLT to be paid based on the contingent event will be subject to a successful application for deferment (see Chapter 12). Valuers should be aware that, although the valuation basis is taken from CGT legislation, the approach to be adopted is not the same as for valuing a "chose in action", which is what unascertained future consideration comprises for CGT purposes.

11 Requirement for Land Transaction Return

A. Introduction

11.1 This chapter deals with circumstances where a Land Transaction Return (LTR), a further LTR, or a supplementary LTR, are required. It also deals with the different computations potentially needed in these circumstances, linked transactions and transactions between spouses and civil partners. The way in which an LTR should be completed, and the penalties for non-compliance, are dealt with in Chapter 13.

B. When a LTR is required

i. Basic provisions

11.2 Section 76(1) of the Finance Act 2003 (FA 2003) states: "In the case of every notifiable transaction the purchaser must deliver a return (a 'land transaction return') to the Inland Revenue before the end of the period of 30 days after the effective date of the transaction."

11.3 The return must include a self-assessment of the tax that, on the basis of the information contained in the return, is chargeable in respect of the transaction. Since the Finance Act 2007 (FA 2007), the tax payable does not have to accompany the LTR, but payment of the amount chargeable must be made within the 30-day period (s 76(3) of FA 2003). Her Majesty's Revenue and Customs (HMRC) also has power, by statutory instrument, to vary or shorten the 30-day period (s 76(2)). It is now a requirement that where an stamp duty land tax (SDLT) 1 is completed and the codes 3 or 4 are entered at question 3, signifying that the transaction comprises of mixed (commercial and residential) or commercial property, then a supplementary return on form SDLT 4 is automatically required.

11.4 While it might be thought a simple matter to decide the effective date, different situations where an LTR is required have their own specific requirements.

ii. Notifiable transactions

11.5 Notifiable transactions are defined in s 77 of FA 2003 as:

a. The grant of a lease for a term of seven years or more where it is granted for chargeable consideration. Except, following the FA 2008 where the annual rent does not exceed £1,000.

b. The grant of a lease for less than seven years if either the chargeable consideration consists of, or includes, a premium in respect of which tax is charged at a rate of 1% or higher, or where the chargeable consideration consists of or includes rent in respect of which tax is chargeable at a rate of 1% or higher (or where the tax would be chargeable in the absence of a relief). (In this respect, one takes into account consideration not covered by the £125,000 threshold for sales, or the £150,000 threshold for commercial transactions. In the case of residential property situated in a designated disadvantaged area (DDA), the applicable threshold is £150,000.)

c. Any other acquisition of a major interest in land is notifiable unless it is exempt from charge under the provisions of Schedule 3 to FA 2003 (see Chapter 1). However, if the transaction is residential and the consideration is less than £1,000, it is considered *de minimis* (s 77(3)(b) of FA 2003). Except, following the Finance Act 2008 (FA 2008) where the chargeable consideration does not exceed £40,000.

d. Any other acquisition of a chargeable interest where the consideration is taxable or would be taxable but for a relief (see Chapter 1, subject to the new thresholds).

iii. Private finance initiative lease and leasebacks

11.6 A private finance initiative (PFI) project involves the sale or lease of land by a public body to a private sector body which, then, leases back the land or underlets the land to the public body. The treatment of PFI projects is set out in Schedule 4 to FA 2003, para 17, which was inserted by the SDLT (Amendment of Schedule 4 to the Finance Act 2003) Regulations (SI 2003/3293). These transactions also include, where appropriate, public private partnerships (PPP). These Regulations provide that where the transaction is a qualifying transaction: "neither the lease-back to the public body nor carrying out of works nor provision of services is chargeable consideration for the transfer or grant of the lease by the qualifying body, or for the transfer of the Surplus Land (in other words tax will generally be charged only on any cash premium or rent paid by the private sector supplier)."

11.7 To qualify as a PFI or a PPP transaction, the transaction must be between a private sector body which is the supplier, and a public sector body, to which the supplies are made. Section 66 of FA 2003 contains an extensive list of those public bodies, including central and local government, health and planning authorities, which qualify for this relief, together with certain bodies concerned with higher or further education contained in para 17 of FA 2003.

11.8 Hence, a primary care trust would be capable of entering into a PFI/PPP transaction, but a group of doctors or dentists, acting together in a partnership, would not constitute a public sector body, and therefore would be unable to qualify for the relief. It is possible for a person to be prescribed to enjoy the relief by Treasury Order.

11.9 All PFI/PPP transactions are notifiable, therefore the private sector supplier would make an LTR at appropriate points, and the public sector body would self-certify on the relevant return.

iv. Assignments

11.10 The assignment of a lease, if deemed to be the grant of the lease at the time of the assignment, would be a notifiable transaction (eg an assignment of the lease held by a charity to an SDLT taxpayer). Also, an assignment where there is consideration chargeable at a rate of 1% or higher, or would be but for any reliefs, is a notifiable transaction (and see para 7.21).

11.11 An agreement for lease is not a notifiable transaction. However, if substantial performance occurs, then it becomes a notifiable transaction and the agreement for lease is effectively treated as if it is the commencement of the lease. Similarly, where an agreement for lease is assigned, if substantial performance occurs, either by the assignor or assignee, then the point of the substantial performance is treated as the commencement of the lease for SDLT purposes.

11.12 Where substantial performance occurs, an LTR is required within 30 days. If any consideration is paid by the lessee to the lessor (or assignee to assignor) following occupation, this sum is potentially taxable. Therefore, if the consideration payable is taxable, then the procedures in paras 11.2 and 11.3 above must be followed. If no taxable consideration is payable, then the self certification procedures must be followed (see Chapter 13). When the actual lease commences, a further return is required on form SDLT 1.

11.13 For the purposes of SDLT, a lease cannot start before the date it is granted following the decision in *Bradshaw* v *Pawley* [1980] 1 EGLR 49, where it was said "no actual term of years can be created until the lease has been executed and so the grant has been made". This is, of course, at variance with the doctrine of substantial performance which is effectively an anti-avoidance measure.

11.14 In practice, an agreement for lease where occupation of the property is granted to the future leaseholder is usually used where works have to be carried out to the satisfaction of the landlord before the granting of the occupational lease. There may be a licence agreement during this period and the payment of a licence fee. Paragraphs 9 and 9A to Schedule 17A of FA 2003 work to give relief in relation to the consideration already taxed under the initial LTR. The mechanism is the same as that adopted in the situation where a tenant holds over, which is dealt with below.

11.15 Where the assignment is that of a lease subject to a relief under s 57A of FA 2003 (sale and leaseback), Parts 1 or 2 of Schedule 7 (group, reconstruction and acquisition relief), s 66 (transfers involving public bodies), Schedule 8 (charities relief), or regulations mentioned in s 123(3) of FA 2003 (interaction with stamp duty relief), then the assignment is treated as if it were a lease granted by the assignor.

11.16 The lease, which in the hands of the assignor was subject to relief, is treated in the hands of the assignee as if it is the grant of a new lease for the remainder of the original term. An LTR is therefore required within 30 days of the assignment (and see para 7.22).

v. Leases for indefinite terms

11.17 Schedule 17A to FA 2003, para 4, deals with the treatment of leases for an indefinite term. Such leases are treated, in the first instance, as if they were a lease for a fixed term of one year. If the lease continues after the end of the initial deemed term of one year, it is then treated as if it were a lease for a fixed term of two years. Thereafter, if the lease then continues after the end of the second deemed term, it is then treated as if it were for a lease for a fixed term of three years, and so on.

11.18 This means that, while at the initial grant of the lease, the transaction may not be notifiable, or require the completion of an LTR, over time the lease and total rents may accumulate to a point where the total consideration becomes taxable. The requirement in Schedule 17A, para 4, is that, when this situation occurs, the purchaser (or tenant) must complete an LTR in the normal way, and return it to HMRC together with the appropriate amount of tax calculated on a self-assessment basis.

Example 11.1

A residential property (not in a DDA) is let for one year at £28,250 per annum (pa).

In year one, the Net Present Value (NPV) of the rent passing is below the threshold of £125, 000.

In years two, three and four, the rent has accumulated to a total of £113,000 and tax is not payable.

However, by the end of the fifth year of the term, the total rent payable will have accumulated, by the end of that year, to some £141,250 which, at a discount rate of 3.5% (YP 5 yrs @ 3.5% = 4.5151), has a NPV of £27,000 × 4.5151 = £127,552.

This therefore brings it above the threshold and an LTR must be returned and SDLT is payable.

11.19 The time at which the leasehold threshold is crossed is the point at which the tax is calculated, and, therefore, any increase in the rate of tax since the initial commencement of the occupation would impact on the amount of SDLT payable, as would any reduction. Therefore, in the above Example 11.1, the LTR would fall to be filed within 30 days of the commencement of the fifth year.

vi Holding over prior to grant of new lease

11.20 There are specific rules in relation to the situation where a tenant "holds over" while negotiating the grant of a new lease. These are set out in para 9 to Schedule 17A of the FA 2003. Paragraph 9A was introduced by s 164 and para 3(1) to Schedule 25 of FA 2006. First, we deal with the situation where a tenant holds over on a year-to-year basis, then where a tenant holds over while in negotiation, leading to the grant of a new lease.

11.21 As shown in Example 11.2 below, Schedule 17A, para 3(2), also provides for the payment of additional tax and the making of an additional LTR, where a fixed term lease comes to an end, with the tenant remaining in occupation and paying rent. Where SDLT was payable on the initial rent during the term then, as the deemed term extends, so further SDLT becomes payable. A tenant in this situation should consider whether it is in their interest, from an SDLT perspective, to seek a further lease rather than pay SDLT on an accumulating basis. This may not be the case in Example 11.2, where a new five-year lease would have a net present value (NPV) below £150,000, but as a six, seven or eight-year lease might have an NPV above the threshold on a cumulative basis, it is clearly beneficial to seek a new lease, particularly where the rent payable remains at a similar level.

Example 11.2

A lease, subject to the Landlord and Tenant Act 1954, is granted and commences on 1 January 2004 at £150,000 per annum for five years.

As the lease is for five years, the NPV is some £677,265 and SDLT payable is £6,773.

LTR is required within 30 days of 1 January 2004.

The tenant remains in occupation at the end of the term paying the same rent. The lease is re-notified and recharged as a lease for six years.

The NPV is now £799,290 (YP 6 yrs @ 3.5% = 5.3286) and SDLT of £1,220 is due (£7,993 less £6,773 paid in 2004).

An LTR is due within 30 days of 1 January 2009 as the effective date of the deemed six-year lease.

If the tenant continues to hold on after 1 January 2010, the process repeats, and so on.

11.22 Where a lease comes to an end by effluction of time it is common that the tenant holds over usually at the existing rent while the terms and conditions of a new lease are agreed. This is an area of some complexity from the perspective of SDLT and the following paragraphs are based on a technical business brief issued by HMRC following the Finance Act 2006 (FA 2006). Where a period of holding over is followed by the granting of a new lease, it is common for the new lease to have as its start date the express end date of the original lease. Thus covering the period of holding over of the original lease. It may also be that the new, increased, rent is payable for this period under the new lease for SDLT return of such a renewable lease cannot start before the date it is granted (per *Bradshaw* v *Pawley* see above). For any period that there is an overlap from the tax previous to the original lease under renewal lease, overlap relief on the rental element of the rent may be claimed. (Para 9 and 9A to Schedule 17A of FA 2003.

11.23 This is despite the fact that the original lease may legally have finished the day before the renewal lease is granted. Where there is an increased rent for the holding over period, if the increased rent during the holding over period is paid for the grant of the new lease, HMRC will treat it as a premium for the new lease and tax it as such. If it is paid for occupation under the original lease, it is taxable as rent under the original lease. The following are examples as set out by HMRC.

Example A

The express term of a business lease came to an end on 13 November 2010, but the tenant continued in occupation while negotiating terms for a renewal lease. The original lease was subject to an annual rent of £150,000. The renewal lease is granted on 1 March 2012, having terms from 1 December 2010 at an increased annual rent of £350,000.

- The original lease was granted or treated as granted prior to 1 December 2003, the period of holding over, as an extension of the original lease, does not need to be notified at all.

- If the increase in rent during the holding over period (£249,315 in total) is paid for the grant of the renewal lease, it will be taxed as a premium for the grant of renewal lease, together with any tax due on the NPV of the renewal lease. Only one LTR has to be made notifying the grant of the renewal lease, if tax is payable or the term of the renewal lease is seven years or more.

- If, however, the increased rent during the holding over period is paid for occupation under the original lease, it is not taxed as a premium. Instead, the new lease will need to be notified and tax paid if tax is payable or the term of the renewal lease is seven years or more. Notification would have to be given of the increased rent under the original lease (by virtue of para 13 to Schedule 17A of FA 2003, treating the rent increase as the grant of the lease. As the duration of this deemed lease is known (15 months) the NPV can be calculated to be some £239,272. (This is the total of £200,000 discounted by 3.5% and £49,315 discounted by 3.5% twice). This deemed grant would not be linked with the original grant but it will be notifiable with tax due of £892, within 30 days of 1 March 2012, this being the effective date of the grant of the renewal lease. As a term end date is one day prior to the effective date, this transaction is to be sent to the Birmingham Stamp Office. Had there been no tax to pay it would not have be a notifiable transaction.

- If the original lease was granted on or after 1 December 2003, the period of holding over, has an extension of the original lease, it will need to be notified if the NPV of the rent for each extra year means that more tax is payable. Therefore, if the lease started on 1 December 2003, the NPV of the original lease would have been £917,181. On 1 December 2010, the lease is treated as extended by one year to 30 November 2011 and the NPV now becomes £1,031,093. Extra tax is due and notification would have to be made by 31 December 2010. On 1 December 2011, the lease is again treated as extended by a further year to November 2012. This results in an NPV of £1,141,152 and a LTR has to be made, as more tax is due.

- If the increased rent during the holding over period of some £249,315 is paid for the grant for the lease, it would be taxed as a premium to be granted for renewal lease, together with any other premium (if any) and on the NPV of the renewal lease. The rent for the first year of the renewal lease (some £350,000) will be reduced by the period of "double counting" of rent of 275 days from 1 March 2012 to 13 November 2012, known as the overlap period. Therefore, the reduction will be £150,000 divided by 365 and multiplied by 275 being some £113,013. The amount to be entered on the NPV calculator for year one of the renewal lease will therefore be £236,987 (£350,000 ñ £113,013). Only one LTR has to be made notifying the grant of the renewal lease, if tax is payable or the term of the renewal lease is seven years or more.

11.24 If a lease is specifically extended during its term this extension is treated as surrender or regrants. SDLT is therefore payable on the NPV on the regranted lease, with the rent element reduced during the overlap period by the rent already taken into account for SDLT purposes.

Example B

The express term of a business lease ends on 30 November 2010. In 2008, the tenant negotiates with the landlord to extend the term of the lease by five years to 30 November 2015. This new lease is granted on 1 December 2008. The rent remains the same at an annual amount of £200,000.

- If the original lease was granted or treated as granted before 1 December 2003, there is no overlap relief for the rent as it has not been subject to SDLT. The new lease for seven years has an NPV of £1,222,908 and should be notified on an LTR.

- If the original lease was granted on or after 1 December 2003, this is subject to SDLT and the rent used to calculate the NPV of the original lease in the overlap period, 1 December 2008 to 30 November 2010, is available for overlap relief. Therefore, the rent to take into account of the new lease in the first and second years is zero, as overlap relief is the same as the rent due. The NPV of the rent for this lease is therefore £842,969.

11.25 It is legally possible for a tenant to enter into a reversionary lease. This is a lease, which takes effect sometime in the future when the current lease expires. The SDLT treatment of these leases depends on whether they are actually granted or whether an agreement of such a lease is entered into. Where a reversionary lease is granted, notification by way of an LTR and any tax are due within the 30 days after the grant. Therefore, where a reversionary lease is granted on 1 December 2005, which has a term from 1 December 2010 to 30 November 2020 at an annual rent of £100,000, the NPV is £831,660. The first years calculation for NPV is £100,000 discounted by 3.5% not £100,000 discounted by 3.5% six times. This is because the calculation of the NPV (per para 3 to Schedule 5 of FA 2003) only starts during the first year of the term of the lease, which in the case of a reversionary lease is the date the lease commences, 1 December 2010 in this example, and not the date when it is granted.

11.26 This compares to the case where an agreement for a reversionary lease is entered into. The agreement is a contract subject to s 44 of FA 2003. Although the tenant may be in occupation to the property, this will be by virtue of the current lease and does not count as substantial performant of the agreement for lease, as occupation is not by virtue of the agreement. When the current original lease determines, the agreement for lease would be substantially performed if the reversionary lease has not already been granted as part of the determination of the original lease. Therefore, SDLT will only be payable at that time and the lease, or substantial performance of the agreement, would become notifiable. Therefore, where an agreement for lease is entered into on 1 December 2005 for a lease to be granted from 1 December 2010 until 30 November 2020 at an annual rent of £100,000. The NPV is again £831,660 but the tax is only payable by 31 December 2010.

11.27 It should be noted that the HMRC automated process for the calculation for NPV is unable to deal with reversionary leases, since the correct start date of the lease is in the future. In these cases, the LTR should be completed in the appropriate way and then sent to the Birmingham Stamp Office with a covering letter explaining the exact circumstances.

vii. Variable or uncertain leases

11.28 Schedule 17A, para 8 of FA 2003, stipulates that, where the rent payable under a lease is variable or uncertain, for example a turnover rent, then, at the end of the first five years of the term of the lease, or the point at which the rent becomes ascertainable, a further return must be made to HMRC (see Example 7.2 in Chapter 7).

11.29 Where a lease is assigned, any responsibility for the making of any supplementary LTR passes to the assignee (Schedule 17A, para 12(1)). This includes the need for a further LTR in cases where a contingency is fulfilled, or in consequence of a later linked transaction, or where rent ceases to be unascertainable. An assignee, therefore, needs to ensure that all relevant information is made available by the assignor so that responsibilities are both identified and capable of being fulfilled.

11.30 Where a lease is varied so that either the term or the rent increase, then a supplementary LTR will be required if more SDLT is payable. Schedule 17A, para 13, states that any increase in rent due to a variation of the lease, is deemed to be the grant of a new lease. Rent increases due to existing contractual arrangements, such as rent reviews, do not come within this provision.

11.31 An increase in the term of a lease is considered a linked transaction (s 108 of FA 2003) and an additional LTR is required (s 81A of FA 2003, and Schedule 17A, para 3).

viii. Surrenders and reduction of rent or lease term

11.32 Where a lease is granted in consideration of the surrender of an existing lease (between the same parties), the surrender is not treated as chargeable consideration for the grant of the new lease. Also, the grant of the new lease does not count as chargeable consideration for the surrender (Schedule 17A, para 16 of FA 2003).

11.33 The release of obligations under the lease in relation to such a surrender, is also not treated as chargeable consideration. These are tenant's obligations, as set out in Schedule 17A, para 10, such as repairing obligations, rental guarantees, penal rents for breach of obligations, etc.

11.34 Schedule 17A, para 15A, states that where a lease is so varied as to reduce the amount of rent payable, the variation is treated as an acquisition of a chargeable interest by the lessee.

11.35 Similarly it states that, where a lease is varied so as to reduce the term, the variation is treated as the acquisition of a chargeable interest by the lessor. In each case the effective date will be the date of the deed of variation and an LTR, together with any SDLT due, must be filed within 30 days.

11.36 While it is apparent what the consideration for the transaction will be where the lessee makes a payment to the landlord for a reduction in rent, or a landlord makes a payment to a tenant to reduce the outstanding term, and to gain an early surrender, if no payment is made by the lessor to the tenant then there will be no consideration to be brought into account.

11.37 What is apparent, however, is that the existence of a condition of the lease to allow for upwards and downwards rent reviews will not be affected by this provision. Nor will an existing condition of the lease allowing for a break clause to be triggered by landlord or tenant. While these are ignored for the purposes of ascertaining the initial length of the lease, they still form part of the contractual terms thereof.

ix. Uncertain or unascertained rents

11.38 Where a lease contains provisions for the adjustment of the rent by a rent review, from a specified date or dates that fall within the first five years of the term, the consideration is calculated at the original effective date on the basis that the contingent sum will be payable. Therefore, where the amount is not known a reasonable estimate is used.

11.39 For increased rents that become payable on review after the fifth year of the lease, they are (generally) ignored for the purposes of the original LTR. For turnover rents and the like, the highest rent payable in any 12-month period of the previous five-year term is used in the recalculation of SDLT (but see also para 11.43 *et seq* below).

11.40 In this situation there are requirements for either the making of a further LTR together with the payment of any additional SDLT due, or the claiming of a refund within the appropriate time limits.

11.41 In the situation of a rent review within the first five years there are three points at which an LTR potentially becomes due (Schedule 17A, paras 7, 7A and 8). First, there is the date at which the rent is ascertained; failing that, the end of the fifth year of the lease. Where the rent is ascertained after the end of the fifth year, a further LTR is required.

11.42 So, in the case of a review in the third year of a lease, it is the point at which the rent is agreed that becomes the effective date, unless still not ascertained at the end of the fifth year of the lease, when an LTR is required, with a further LTR on the later ascertainment of the rent.

Example 11.3

A lease is granted for five years, with rent increases to be based on turnover.

An initial reasonable estimate (Schedule 17A, para 7) is included in an LTR.

After five years a revised LTR is submitted based on highest rent paid in any 12-month period.

Additional SDLT is payable (or refundable) depending on estimated and actual rent ascertained.

Example 11.4

A lease is granted, as in Example 11.3, but for seven years.

An initial reasonable estimate for first five years is made, from which is determined the highest rent likely to be payable in first five years.

The LTR includes NPV calculated for seven-year term.

After five years, an amended LTR is required with revised NPV calculation submitted.

Rent is determined from highest amount actually paid in any 12-month period in first five years.

SDLT is payable on seven-year lease calculation, with credit for original SDLT paid, or a refund is made.

Example 11.5

A seven-year lease with rent review at year four is granted on 1 January 2005.
Initial rent is £100,000 per annum.

Estimated rent at year four is £125,000.
Rent agreed on review £150,000.

(a) **Assume that the review was concluded on 31 December 2008**
The following LTRs would be required.

By 31 January 2004, with NPV based on £100,000 (years 1–4) and estimated rent of £125,000 (years 5–7).

By 30 January 2009, an LTR with NPV based on £100,000 (years 1–4) and £150,000 (years 5–7).

(b) **Assume that the review was concluded on 1 July 2008**
The following LTRs would be required.

By 31 January 2004, with NPV of £100,000 (years 1–4) and estimated rent of £125,000 (years 5–7).

By 31 July 2008, an LTR with NPV based on £100,000 (years 1–4) and £150,000 (years 5–7).

(c) **Assume the review was concluded on 1st July 2009 after dispute**
An estimate on 30 January 2009 would be required (say £140,000) as the five-year point is reached at 31 December 2008. The following LTRs would be required.

By 31 January 2004, with NPV based on £100,000 (years 1–4) and an estimated rent of £125,000 (years 5–7).

By 30 January 2009, with NPV based on £100,000 (years 1–4) and a re-estimated rent of £140,000 (years 5–7).

By 31 July 2009, with NPV based on £100,000 (years 1–4) and known rent of £150,000 (years 5–7).

X. *Abnormal increases in rent*

11.43 Chapter 7, Schedule 17A, paras 14 and 15 of FA 2003 set out the conditions under which a rental increase can be considered abnormal. Basically, the date of review in relation to these provisions falls at or after the end of the fifth year of the lease. These have been subject to significant change introduced by s 164 and para 8(1) to Schedule 25 of FA 2006 which reduced the six steps envisaged in the original legislation to three.

11.44 The question the taxpayer must ask at the point of review is whether or not the abnormal increase provisions are in point. These provisions will not affect rent reviews prior to December 2008. The legislation originally provided that, where the annual rent payable after year five increased by more than 5% plus Retail Prices Index (RPI), over and above the rent used in the original LTR, then such increases were regarded as abnormal and would be treated (deemed) as the grant of a new lease. It can be seen that if the RPI increased at, say, 3% per annum then, the compound effect of rent plus RPI would lead to an increase of 50% over, say, five years and would trigger the provisions.

11.45 The provisions of para 15 to Schedule 17A of FA 2003 have been modified by the introduction of a formula which requires the application of three steps.

- Find the start date.
- Find the number of whole years between the start date and the date on which the new rent first becomes payable.
- Test the new rent against the formula $\frac{R \times Y}{5}$

  where R is the rent previously taxed and where Y is the number of whole years.

The excess rent must be greater than the product of the formula, this is a slightly complex way of stating if the rent increases by more than 100% then it will be treated as abnormal.

11.46 Assume a lease for 21 years is entered into on 1 January 2004 (the start date), subject to reviews at three yearly intervals. At the first review in January 2007 no additional (abnormal increase) LTR is required. However, in January 2009 the rent passing, as agreed under the review as at 2007, must be tested to see if it can be considered abnormal. Then, following each review in 2010, 2013, 2016, 2019 and 2022, the rent must be retested to see if, with reference to the original LTR submitted in January 2004, or to the last occasion when there was an abnormal increase, the rent increases can be considered and a fresh LTR may be required with additional SDLT.

11.47 The LTRs generally required, however, in the context of this lease, would follow those in Example 11.5 above, the abnormal rent increase calculations providing an additional requirement.

11.48 While these provisions, from a valuation standpoint, seem to be more logical than the initial ones as they no longer assume rents should move within an RPI plus a set percentage driven framework. In practical terms they require the re-evaluation of rental payments at regular stages throughout the term of the lease. Therefore, beneficial variations to a locality, which change value patterns in terms of rents are not, apparently, to be taken into account, only increases in rental value. Effectively, this allows for the additional payment of SDLT if market forces drive rents above a ceiling.

11.49 The deemed grant of the new lease is taken as being made on the date the first increased rent becomes payable, and the "new" lease is taken as being for the, then, unexpired term of the lease. Therefore, in para 11.44 above, if the review of 2010 brought the abnormal rent increase provisions into play, then the deemed lease would be for a term of 15 years.

11.50 It should also be appreciated that, if the rent increases again during the course of the "deemed" lease, and it can be judged abnormal by reference to the rent used in the calculation of the NPV in the most recently returned LTR, then a further lease can be deemed to be granted. There is no additional five-year rule for the deemed leases as they are past the point of the initial five years of the lease, the transactions being treated as linked transactions to the originating lease.

11.51 A further consideration is the meaning of Schedule 17A, para 14(5). It refers to the date on which the increase in rent first becomes payable. This will, possibly, unless the lease otherwise provides, be the date of the review or the date on which the rent increase is agreed. Normally, a reviewed rent becomes payable from the quarter day following the date of the new rent being determined, but calculated from the review date. In this situation, the specific provisions of the lease define the effective date.

11.52 This leads to the interesting situation that a taxpayer may need to make an LTR within 30 days following the review date after the fifth year of the term, where they believe that the review could lead to the creation of a deemed lease under the abnormal increase provisions after estimating the amount of rent payable on review.

11.53 As a rent review, seeking a significant uplift in rent may lead to extended negotiations and even litigation, the level to which the rent may rise must be estimated. It is also unlikely that deferment under s 90 of FA 2003 of tax may be sought as the rent, while "uncertain", falls to be paid prior to the date of the LTR (from the first quarter day) rather than at least six months hence. In addition to the abnormal rent LTR, any other LTR requirements, as set out above, also apply.

xi. Right to buy, shared ownership transactions, shared ownership leases, etc

11.54 Section 70 of FA 2003 introduces Schedule 9 to FA 2003. This schedule is effectively split into three parts and deals with:

a. relief for right to buy (RTB) transactions (para 1)
b. relief for shared ownership leases (paras 2–5)
c. relief for rent to mortgage and rent to loan transactions (para 6).

11.55 The reliefs all impact on the way in which an LTR is completed and also on the number of LTRs required in certain circumstances. The definitions of qualifying transactions are dealt with more fully in Chapter 17.

11.56 The FA 2007 extended SDLT relief for affordable housing, set out below, to shared ownership trusts.

11.57 In the case of RTB transactions, the relief operates by excluding any contingent consideration which may occur. This is effected by disapplying some provisions relating to chargeable consideration (s 51 of FA 2003). Therefore, chargeable consideration only applies to the discounted amount payable by the purchaser at the point of initial purchase, subject, of course, to the appropriate thresholds for residential property (see section B, paras 17.4 to 17.10 in Chapter 17).

11.58 Shared ownership leases consist of the landlord granting the tenant a long lease for a premium equivalent to the share acquired, with rent payable on the remaining portion. Over time the tenant may acquire the entirety of the property, in some cases the freehold reversion, in others a long leasehold interest.

11.59 The tenant may return an LTR on one of two bases, either on the separate parts as each part is purchased, or on the basis of what SDLT would be payable if the entirety was purchased at the commencement of the lease or at any stage thereafter (see also section C, paras 17.11 to 17.21 of Chapter 17). Note also that a different basis of valuation applies (see also Chapter 10).

11.60 Where the lease meets the statutory requirements, there are now two further conditions to be met. First, the lease must be granted by a qualifying body namely:

- a local housing authority
- a housing association
- a housing action trust
- the Northern Ireland Housing Executive
- the Commission for the New Towns
- a development corporation.

Alternatively, the lease is granted in pursuance of a preserved RTB. This is defined as where:

- the lessor is a person against whom the RTB (under Part 5 of Housing Act 1985) is excercisable by virtue of s 171A of that Act
- the lessee is (or lessees are) the qualifying persons for the purposes of the preserved RTB
- the lease is of a dwelling, that is a qualifying dwelling-house in relation to the lessee.

11.61 The lessee must elect for SDLT to be charged in accordance with para 4 to Schedule 9 of FA 2003. The election must be made on the LTR submitted in respect of the grant of the lease, or an amendment to that return. The election is irrevocable, that is, once made it cannot be changed, even by an amendment, or further amendment, to the LTR.

11.62 Where the tenant pays SDLT on the whole at the commencement of the lease, then no further SDLT is payable, although self-certification will be required on acquisition of the whole.

11.63 An advantage to the entirety approach will be if the value at commencement falls below the residential threshold, particularly in a DDA.

11.64 With regard to reversions, para 3 to Schedule 9 of FA 2003 applies. Where:

- a valid election has been made under para 2 to Schedule 9 of FA 2003

- any SDLT has been paid.

Then, the subsequent transfer of the reversion under the lease agreement is exempt from charge.

11.65 Where no SDLT is due on the shared ownership transaction, it will be necessary for the lessee to make an election under para 2 to Schedule 9 of FA 2003 to ensure the transfer of the reversion is exempt from charge. No SDLT is due, for example, if the consideration falls within the 0% SDLT threshold or disadvantaged areas relief is claimed.

11.66 The reliefs for rent to mortgage give relief for certain purchases under the Housing Act 1985. In the case of rent to loan, SDLT is limited to that payable if the consideration was calculated under s 62 of Housing (Scotland) Act 1987 rather than the actual purchase price (see paras 17.22 to 17.25 in Chapter 17).

xii. Linked transactions

11.67 Linked transactions are defined in s 108 of the FA 2003, which states transactions are "linked" for the purposes of this part of this Act if they form part of a single scheme, arrangement or series of transactions between the same vendor and purchaser, or in either case, persons connected with them (see Chapter 16).

11.68 Where two or more transactions have the same effective date they may be notified using a single LTR. The rate of tax being defined by the sum of the chargeable considerations paid. This procedure is the same as that for "joint purchasers" as set out in s 103 to FA 2003.

11.69 There is a certain amount of confusion surrounding the area of linked transactions, mainly stemming from HMRC's *Stamp Duty Land Tax Manual* at SDLTM 30100, where it also states:

> It is a question of fact whether or not transactions are linked. A purchaser will need to make a full examination of all the circumstances leading to the transactions before completing their LTR. Just because two transactions are between the same purchaser and seller does not necessarily mean they are linked. The transactions will be linked however if they are part of the same deal.

11.70 First, let us look at what comprises a linked transaction. If a purchaser acquires an entire property by way of two or more transactions, at the same time, from the same vendor, say, a flat over a shop, this would clearly be a linked transaction. Where a purchaser acquires, say, three flats from a developer and recieves a discount, perhaps by buying off plan, this would clearly form part of the same "deal" and be a linked transaction. The legislation and the SDLT manuals are aimed at preventing the artificial fragmentation of transactions to reduce the amount of tax properly payable.

11.71 However, what is the situation where a purchaser acquires two flats, individually, in the same block from the same vendor and recieves no discount? HMRC's view is that if the transactions have the same effective date they are linked, or at least this is what its initial view was, although the wording of s 108, together with the wording of the *Stamp Duty Land Tax Manual* (which does not have force of law) would indicate the possibility of a different view. Current experience appears to indicate that where two transactions occur at the same time, but are conducted as if they were individual transactions, then they will not be treated as being linked.

11.72 Similarly, if a husband and wife purchased two individual flats from the same developer, independently, but at the same date, by virtue of being connected persons these transactions would be apparently linked. However, this view would appear to be capable of a different interpretation.

11.73 The key here is probably the effective date, together with the parties intentions at the point of the first transaction, where individual properties are purchased over time without the intention of the transactions being linked as part of a series, then factually they would be separate transactions. For example, if one spouse purchased a flat and the other spouse subsequently purchased another, in the same block, then it would be a matter of fact that the transactions were not linked. Similarly, if an investor purchased a shop property and subsequently purchased the living accommodation above, perhaps on it falling vacant, it would be hard to see this as a linked transaction unless there was some form of agreement in place between vendor and purchaser at the point of the initial transaction.

11.74 The taxable amount, and the applicable rate, is judged from the sum of the chargeable consideration. HMRC's penalty notice forms use the word total consideration, in connection with linked transactions. This has led some HMRC officers, mistakenly, to argue that non-chargeable consideration can be aggregated together with chargeable consideration in setting the applicable rate. This view is wholly wrong as only chargeable consideration is taxable to SDLT, and therefore capable of aggregation.

xiii. Transactions between spouses and civil partners

11.75 There is no automatic relief for transactions between spouses or civil partners, even when associated with marriage, divorce and civil partnership. Various different scenarios may arise as set out below. However, as SDLT is charged where an interest in land is transferred for consideration, it should be borne in mind that consideration will include:

- any cash payment
- any assumption of liability to pay a mortgage
- the liability assumed is taken to be a proportion of the outstanding mortgage corresponding to the proportion of the share of the property acquired.

11.76 For example, say a house is valued at £180,000. The transferring partner has equity of £90,000 and there is an outstanding mortgage of £90,000. The transferee partner pays a cash sum equivalent to half the equity and acquires a 50% share in the property. The consideration is therefore the cash payment of £45,000 plus 50% of the outstanding mortgage totalling together some £90,000. As this is below the SDLT threshold of £125,000 there will be no tax to pay. However, details of the transaction must be returned using an LTR. For this purpose, joint tenants are treated as if they owned 50% each of the property, rather than an undivided share of the whole.

11.77 The situation is different where a couple are divorcing, dissolving a civil partnership or splitting up and wish to transfer the property from their joint names into the name of one partner. Where such a transaction is effected in pursuance of a court order or an agreement between the parties in connection with divorce, nullity of marriage, judicial separation, or the dissolution of a civil partnership, it is exempt (Schedule 3 to FA 2003).

11.78 In this case, the transaction can be self-certified to the Land Registry using form SDLT 60 and no LTR is required. Otherwise, SDLT will be charged as set out above.

11.79 For example, a house is valued at £350,000, the partners have an equity of £250,000 and there is an outstanding mortgage of £100,000. The transferree partner pays a cash sum equivalent to 50% of the equity and acquires sole ownership of the property. The consideration is therefore the cash payment of £125,000 plus 50% of the outstanding mortgage totalling some £175.000. Therefore, tax is payable at 1% in the sum of £1,750. An LTR must be made in the usual way.

xiv. Zero-carbon houses and flats

11.80 This relief is effective from 1 October 2007, s 52B FA 2003, and runs until 30 September 2012, and has been extended to include flats in the Finance Act 2008 (FA 2008). It is time limited to encourage the provision of zero-carbon "homes" whose provision will be mandatory by 2016.

11.81 The relief applies only to "new" property at first point of sale to an end user. Relief is claimed by ticking the appropriate box on form SDLT 1. Relief can be claimed in respect of properties with a value in excess of £500,000 by making a deduction from the SDLT payable of some £15,000. Under self-assessment regulations, the claimant has to prove that the property qualifies under Standard Assessment Procedure (SAP) 17A Building Regulations 2000 (SI 2000/2531).

11.82 This requires the provision of a certificate provided by an accredited assessor, the certification process being overseen in England and Wales by the Department for Communities and Local Government, in Northern Ireland by the Departments of Finance and Personnel or Social Development.

xv. Surrender of and forfeitures of leases

11.83 Where a lease is surrendered for little or no consideration and is either registered or there is an express deed of surrender, then a form SDLT 60 is required. In other circumstances which do not amount to an express surrender of and regrant of a lease, no notification is required.

11.84 Similarly, forfeiture of leases because of a breach of covenant, or non-payment of rent, may not constitute a chargeable transaction. This is a complex area where specialist advice may

well be required. The SDLT manuals suggest that "none of these situations should trigger a charge to SDLT", which indicates a degree of uncertainty which is of little assistance to the taxpayer who is responsible under the self assessment provisions.

12 Deferment of Payment of Tax

A. Application for relief under s 90 of Finance Act 2003

12.1 Under certain circumstances, as set out in paras 10 to 28 of the Stamp Duty Land Tax (Administration) Regulations 2003 (SI 2003/2837) (the Regulations), an application for the deferment of stamp duty land tax (SDLT) may be made in accordance with Part 4 of the Regulations where the whole or partial consideration:

a. is contingent or uncertain as defined in s 51 of Finance Act 2003 (FA 2003)
b. it becomes payable, or may become payable, more than six months after the effective date of the transaction.

12.2 The application must be sent to:

Birmingham Stamp office
9th Floor, City Centre House
30 Union Street
Birmingham B2 4AR
DX: 15001 Birmingham 1
Fax: 0121 643 8381

12.3 Any application must be in writing, be marked SDLT Deferment Applications, and set out:

a. the identity of the purchaser
b. the location of the land involved

c. the nature of the contingency or uncertain event
d. the amount of consideration for which deferment is sought
e. as full details of the times of expected payments as it is possible to give
f. a reasoned opinion as to when this part of consideration will cease to be contingent or can be ascertained
g. a calculation of the SDLT potentially payable on the total of the actual contingent or uncertain consideration
h. a calculation of the SDLT in respect of which the application to defer refers.

Her Majesty's Revenue and Customs (HMRC) may ask for further information. However, HMRC must say why this information is required, and allow not less than 30 days for a response. If the information is not provided in the time allowed, the application may be refused. Once the application for deferral is made, the payment of SDLT claimed as deferred is suspended until a decision on the claim has been made.

12.4 Any application must be made on or before the filing date of the Land Transaction Return (LTR).

B. Applicant's obligations

12.5 The purchaser is still obliged to pay SDLT on consideration which:

a. has already been paid when the application is made
b. will become payable within six months of the effective date of the transaction
c. is not contingent
d. is known at the time when the application is made.

12.6 Deferment is not available on consideration that is ascertainable but not yet ascertained. This is because, for the return to be made in the basis set out in s 51(3) of FA 2003, the amount must depend on the outcome of uncertain future events. In this case, the LTR must be completed on the basis of the best estimate of each purchase price and an amendment, under Schedule 10 to FA 2003, para 6, should be made on the final consideration when ascertained.

C. Special rules for works and/or services

12.7 Schedule 4 to FA 2003, para 10/11, sets out special rules. If the works (para 10) or provision of services (para 11) are expected to take less than six months to complete, SDLT must be paid within 30 days of completion in the normal way.

12.8 If the works or services are expected to take more than six months, the application must set out payment dates at intervals of not less than six months, with a fixed payment of SDLT to be made within 30 days of completion of the particular works, stage by stage.

12.9 Payment will be based on the value of the works or services. An application to vary the payment schedule may be made near to the end of the works where payment is due but works are mainly complete. Any application must be made in writing. This regulation is to cover the situation where the additional payment would be required a short period after the next one due, and the payments could be rolled into a final one.

D. Decisions by HMRC on applications

12.10 A notice must be given in writing where the application is accepted or refused. Where the HMRC accept the application, the notice must set out the terms and specify:

a. any tax payable in accordance with the LTR
b. the nature and dates of any relevant events
c. how any tax payable is to be calculated at the various dates
d. that tax is payable within 30 days of a relevant event in accordance with Part 4 of the Regulations.

12.11 Each payment must be accompanied by a return, these returns are treated as LTR and the rules in Schedule 10 to FA 2003 apply. Provisions for the returns are set out in SDLT (Administration) Regulations 2003 (SI 2003/2837).

12.12 Payments are made on the due dates until all the tax liability is discharged. At each stage the chargeable consideration for the whole transaction is reviewed to ensure the correct rate of tax is being paid. If the amount of tax payable reduces because the consideration becomes less, then overpaid tax plus interest may be reclaimed. If the application does not contain full and proper information it shall have no effect, if any information proves either incorrect or relevant facts or circumstances have not been disclosed it shall also have no effect.

12.13 Where an application is refused, the notice must set out:

a. the grounds for refusal
b. the total SDLT payable in consequence.

E. Grounds on which application may be refused

12.14 Grounds for refusal include:

a. the conditions of s 90(1) of FA 2003 are not met
b. the application does not comply with regulations 12 or 14 (ie incorrect format or insufficient information)
c. there are tax avoidance arrangements in relation to the transaction (regulation 18)
d. the application is incorrect or incomplete
e. information required under regulation 14 was not provided in reasonable time (see para 12.3).

F. Appeals

12.15 Appeal lies with the General or Special Commissioners.

12.16 Any appeal must, under regulation 20:

a. be in writing
b. be made within 30 days of the issue of the refusal
c. be to the officer of the board who gave the refusal
d. the appeal must set out the appropriate grounds on which it is made.

(However, it should be noted that the Commissioners may allow further grounds to be put forward and considered on behalf of the appellant if the omissions are neither deliberate nor unreasonable.)

G. Settlement by agreement

12.17 Before any appeal made under regulation 20 is determined, the appellant and HMRC may agree that:

a. it should be upheld without variation
b. it should be varied in a particular manner
c. it should be discharged or cancelled.

12.18 Following any agreement, the same consequences shall follow, as if the commissioners had determined the appeal and had either upheld the decision without variation, or varied it in the manner agreed or discharged, or cancelled it as the case may be.

12.19 No agreement can be reached if, within 30 days from when the agreement was made, the appellant gives notice in writing that they wish to withdraw from the agreement.

12.20 This can only apply where the terms agreed have still to be confirmed by notice in writing, given by either HMRC to the appellant or by the appellant to HMRC. In these cases, the date of the notice of confirmation is taken as being the time at which the agreement was reached.

12.21 However, where the appellant notifies HMRC that they do not wish to proceed with the appeal, and HMRC do not, within 30 days, give notice indicating that it is willing that the appeal should be withdrawn, then the matter shall be treated as if an agreement had been reached in which the decision under appeal should be upheld without variation.

12.22 If it is agreed that the payment of any amount of SDLT should be postponed, pending the termination of the appeal, then matters shall proceed as if the Commissioners had made a direction to that effect.

H. Directions by Commissioners

12.23 It is open to the Commissioners, having received an application in writing by the appellant, to postpone payment of SDLT until after the appeal has been determined. This application must be made within 30 days from which the notice or decision to refuse the application has been issued, and should state the amount of tax to be postponed.

12.24 On determination of the appeal, SDLT becomes payable once HMRC issue a notice of the total amount payable in accordance with the determination. Alternatively, at that point, HMRC should repay any overpaid tax.

I. Applications having no effect

12.25 Under certain circumstances, an application which has been accepted by HMRC may have no effect. This is both for the purposes of Part 4 of FA 2003 and the Regulations. The circumstances are:

a. if the application turns out to have contained false or misleading information

b. any facts or circumstances relevant to the application are not, or have not been, disclosed to HMRC. If the facts and circumstances relevant to the application change for these reasons, it shall be assumed that the application accepted by HMRC has no effect and, therefore, either a further application is required or SDLT on the transaction must be paid.

Administration of Stamp Duty Land Tax

13

A. Land transaction returns

13.1 The purchaser must deliver a Land Transaction Return (LTR) to Her Majesty's Revenue and Customs (HMRC) within 30 days of the effective date of a notifiable transaction (s 76(1) of Finance Act 2003 (FA 2003)), also referred to as the filing date. (For the definition of notifiable transactions see also Chapter 11.)

13.2 Since the introduction of stamp duty land tax (SDLT) on 1 December 2003, the administration of SDLT has been more codified with the introduction of a range of forms, guidance and additional returns. These returns can be submitted in a number of ways, namely:

- by way of online services
- using an HMRC CD-ROM
- on green and white paper forms
- on black and white 2D bar-coded forms.

13.3 The primary forms of return and relevant guidance notes are:

- SDLT 1: Land Transaction Return
- SDLT 2: Supplementary form, where there are two or more sellers and/or two or more buyers
- SDLT 3: Supplementary form, where land is involved and more space than that provided on the SDLT 1 is needed
- SDLT 4: Supplementary form, for complex commercial transactions and leases

- SDLT 46: Appeal against Penalty Determinations (since April 2006 this "new" form must be sent to the Newcastle Stamp Office)
- SDLT 60: Certification that no LTR is required for a land transaction
- SDLT 60: Guidance notes on how to complete SDLT 1 and other forms
- SDLT 68: Guidance on how to complete an SDLT 60.

13.4 The returned LTR, known as an SDLT 1, an example of which is included at Appendix 1, must, in accordance with Schedule 10 to FA 2003 and the SDLT (Administration) Regulations (SI 2003/2837) (the Regulations):

a. contain a calculation of the SDLT due
b. be in the form set out by the Regulations (Schedule 10, to FA 2003, para 1) (it should be noted that HMRC retain the right to vary the form of return)
c. contain all required information (see also para 13.5)
d. include a declaration as to the accuracy of the information provided.

Where on an SDLT 1 question 1 is answered using code 02 or 03, mixed or non-residential property, a supplementary form SDLT 4 must also be submitted, an example of this is contained at Appendix 1.

13.5 Prior to the Budget 2007, the LTR had to be accompanied by payment of the tax payable (subject to a 12-day period of grace where a return was made electronically). Since the amendment of s 80 Finance Act 2007 (FA 2007), it is not necessary for the return and payment to be delivered together, however both must be returned or paid no later than the filing date. The LTR must always be signed by the purchaser or purchasers personally, not by an agent or representative. The exceptions to this is a holder of a power of attorney, together with those acting in a representative capacity, such as executors, receivers, representatives of incapacitated persons, and parents and guardians of minors who may also be required to make a suitable declaration to this effect (s 81B of FA 2003). It should also be noted that payment by cheque of SDLT is subject to separate regulation (s 92 of FA 2003).

13.6 Section 106(1) of FA 2003 allows such representatives to access sufficient funds to cover SDLT liability. In most cases, the section also provides that the actual taxpayers are liable for the SDLT payable, and the representative is indemnified by them. However, in the case of parents and guardians, they are deemed to be responsible for any shortfall in funds to pay outstanding SDLT.

13.7 In the case of a company, any properly nominated officer of the company may sign the LTR and, in the case of a partnership, a representative partner may sign and accept joint and several liabilities on behalf of the partnership as a whole. However, if the transaction is subsequently successfully challenged by HMRC, only those partners who were partners at the effective date are liable for unpaid tax or penalties.

13.8 Due to the use of electronic scanning, LTR forms are required to be completed in black ink. Each form is unique and not capable of duplication.

13.9 The required information, much of which can be obtained from the explanatory leaflets where not already known, includes:

a. where the purchaser is an individual, their National Insurance (NI) number (this condition is only relevant to those who have UK NI numbers)

b. the National Land Property Gazetteer Unique Property Reference Number, if one exists in relation to the particular property

c. for development land, agricultural land and garden plots, suitable plans are to be provided.

13.10 Notifiable transactions requiring an LTR, under s 77 of FA 2003 and relevant schedules, are dealt with in detail in Chapter 11. In summary, they are:

a. the grant of a lease for more than seven years for chargeable (taxable) consideration

b. the grant of a lease for less than seven years if for chargeable (taxable) consideration, or would be so but for relief. In this case, to be taxable, the consideration must exceed any relevant threshold

c. any other acquisition of a major interest (see Definitions for explanation) unless exempt by Schedule 3 to FA 2003

d. any other acquisition of a chargeable (taxable) interest if it is taxable, or would be but for relief

e. all private finance initative transactions.

B. Compliance

13.11 An LTR is to be fully completed by the purchaser with all requisite information. In this context, all requisite information is considered to be that which is required for proper assessment of any SDLT due (Schedule 10 to FA 2003 and the Regulations — and see para 13.2). In this respect, HMRC also draws support from the case of *Langham (Inspector of Taxes)* v *Veltema* [2004] EWCA Civ 193 in which the Court of Appeal considered the responsibilities of the taxpayer under the self assessment regime in relation to the provision of full and proper information. The matter of enquiries and discovery assessments is dealt with in Chapter 14 hereafter. The provisions are contained in Part 3 of the Regulations and the various forms for an LTR are set out in Schedule 2 thereof. Sections 97 to 99 of FA 2003 also allow the Lord Chancellor to amend regulations in this respect and to deal with the circumstances in which tax is repaid, after being paid under a proposed regulation not subsequently enacted.

13.12 If a purchaser fails to deliver a return by the filing date (30 days after the effective date of the transaction), there may be a penalty levied of £100 if the return is delivered within three months after the filing date, rising to £200 in any other case (Schedule 10 to FA 2003, para 3). Where the LTR has still not been filed within 12 months of the due date, there may be a further tax related penalty of up to 100% of the amount properly payable (Schedule 10, para 4). Interest may also be payable on unpaid penalties (s 88 of FA 2003).

13.13 Taxpayers who deliver late returns often appeal when served with a penalty notice on the grounds of "reasonable excuse". HMRC's view is that any reasonable excuse depends on some exceptional event beyond the purchaser's control. HMRC cites three circumstances in illustration, the return is lost or delayed in the post, the taxpayer's advisor (with responsibility for submitting the return) is taken seriously ill, or the advisor dies.

13.14 It should be noted that the SDLT penalties in general are cumulative and could amalgamate to a significant level.

However, by concession in the Budget of 2004, the Chancellor stated that the penalties should not cumulatively exceed 100% of the amount of tax properly charged. This is now included in s 99(2A) of FA 2003 which provides that the liability shall not exceed the greatest penalty out of the several penalties. The *Stamp Duty Land Tax Manual* 86600 and 86620 refer to a reduction of up to 40% of the penalty if the taxpayer co-operates. Interest may still accumulate on unpaid tax and penalties to raise the total amount payable above the 100% threshold. In the case of all penalties for SDLT they do not form an allowable deduction for any other tax.

13.15 If HMRC discovers, or believes, that an LTR should be filed, then it may serve a notice on the purchaser requiring the purchaser to do so. The notice must specify the relevant transaction and a compliance period of at least 30 days from the date of the notice (Schedule 10 to the FA 2003, para 5).

13.16 If the purchaser does not comply with the notice, HMRC may apply to the General or Special Commissioners for an order to allow the imposing of a daily penalty not exceeding £60 until compliance is achieved. This penalty is additional to any other penalties levied but subject to the overall limitation (Schedule 10, para 5). In practice, the purchaser would either comply with the notice, or appeal against it to the Commissioners and, when the matter is subject to an appeal, the purchaser, or alleged purchaser, would doubtless apply for deferment of tax potentially payable until the appeal is settled.

13.17 It is possible to amend an LTR subsequent to it being filed. Schedule 10 to FA 2003, para 6, sets out two basic conditions in this respect.

- The amended return must be in a form and contain such information as HMRC may require.
- Normally, such an amended return must be made within the 12 months following the filing date.

13.18 Where an LTR contains obvious mistakes or omissions, HMRC has the right to amend the LTR. However, this must be done within nine months of the day on which the return is delivered, or of the day on which any amendment is made (if the correction is made to that amendment) (Schedule 10 to the FA 2003, para 7).

13.19 In common with other compliance procedures with SDLT, HMRC must issue an appropriate notice to the purchaser. An amendment proposed by HMRC has no affect if the purchaser corrects the return appropriately themselves or, if HMRC are out of time, the purchaser may reject the amendment by giving notice to the appropriate officer. This, however, has to be carried out within three months of the notice being issued by HMRC (Schedule 10, para 7).

13.20 In relation to an incorrect return by a purchaser which is either fraudulent or negligent, or where the purchaser discovers that a submitted LTR is incorrect, if the purchaser does not remedy it within a reasonable period, they are then liable to a penalty of 100% of tax understated on the return (Schedule 10 to the FA 2003, para 8). For example, if the true amount of tax payable was £1,500, but the return had stated it as being £1,000, then the penalty would relate to the understated amount of £500. This penalty could thereafter be combined with other penalties, if appropriate, as the penalty ceiling in the particular case would be an additional £1,500, ie 100% of the tax properly payable.

13.21 The delivery of an LTR to HMRC, together with the appropriate amount of tax, should not be confused with the process of registering a land transaction with the Land Registry. In order to register a land transaction, it is necessary for the purchaser to be able to produce a certificate issued by HMRC following receipt of a properly completed LTR, together with the appropriate tax payable (s 79(3) of FA 2003).

13.22 In the circumstance where an LTR is not required in respect of a transaction then, in order to satisfy the Land Registry's requirements, the purchaser must issue a self certificate declaring that no return is required to be filed (s 79(3)(b) of FA 2003).

13.23 There are a number of circumstances where an additional SDLT 1 must be submitted, where an agreement or contract was previously notified because of "substantial performance" (see chapter 1) needs to be modified on "completion". A substantially performed agreement for lease is treated as a lease. The grant of the lease is then, for SDLT purposes, another lease, which is linked for threshold purposes (but not for the purposes of para 5 to Schedule 17A to FA 2003). So, for SDLT purposes there are two leases, each of which needs an SDLT 1 and ADLT 4 unless the actual lease can be self-certified (see below). The actual lease benefits from a relief (para 9 to Schedule 17A of FA 2003) which reduces the rent, and it is the reduced figure that is used in the net present value (NPV) calculation of the actual lease. It should be noted that the para 9 relief works by reducing the rent on the new (or in this case actual) lease, not by giving credit for SDLT previously paid.

13.24 If the threshold linkage gives rise to further tax payable on the substantially performed agreement for lease, then a further return must be made to the Birmingham Stamp Office, in the form of a covering letter until forms are produced for further returns. This is because of the unique nature of each SDLT 1.

C. Self-certification

13.25 The process of self-certification is very similar to that adopted in returning an LTR, but there are a number of important differences. The appropriate form is SDLT 60, which is available from The Stationery Office (TSO) (formally Her Majesty's Stationery Office) and from local land registry offices. Unlike the LTR (Form SDLT 1), a self-certificate is not a unique form and it is acceptable to use photocopies of a blank. The return is also made direct to the land registry.

13.26 Direct similarities with an LTR.

- It must be in the prescribed form (s 79 of FA 2003).
- It must include a declaration by the purchaser or purchasers that the certificate is correct and complete. Prior to the Finance Act 2007 (FA 2007), this declaration had to be personally signed by the purchaser. Section 81 of FA 2007 makes provision for the purchaser or the purchaser's authorised agent to either sign and or make the appropriate declaration, specific regulations in this regard are to be made following the Finance Act 2008.
- In the same way as set out in para 13.17 above, there is a potential penalty of up to 100% of the tax properly payable in the cases of fraud, negligence, or error that are discovered by the purchaser and not expeditiously corrected. An example of correction would be the filing of an LTR, and the payment of the appropriate tax where it is discovered by a purchaser that self-certification was inappropriate.

13.27 There are no penalties in place for failure to self-certificate, in the same way as there are penalties for failing to return an appropriate LTR. However, without self-certification or the appropriate certificate issued by HMRC (see para 13.18 above), the purchaser will not be able to register the transaction and thereby prove the proper title to the relevant land.

13.28 Under the provisions of s 79(6) of FA 2003, HMRC retains the right to inspect any certificate or self-certificate produced to the Land Registry, and the Land Registry may provide HMRC with any other necessary information or facilities so that it may verify that the statutory requirements have been complied with.

D. Exceptions and supplementary returns

13.29 Sections 44, 45 and 45A of FA 2003 provide circumstances in which no certificate is required to register a contract for a land transaction. These are circumstances where transactions are entered into which are similar to the subsale transactions often entered into under stamp duty. This matter is dealt with in more detail in Chapter 11. However, where the appropriate circumstances apply, the purchaser or "person" is not regarded as having entered into a land transaction. This can apply both to a land transaction and to a transfer of rights.

13.30 Section 44A of FA 2003 effectively gives an example. Where a contract is entered into under which a chargeable interest is conveyed by one party to the contract (A) at the direction or request of the other (B), to a person (C), who is not a party to the contract, or either to C or to B, then B is not regarded as entering into a land transaction by reason of entering into the contract. However, as pointed out in Chapter 1, if during the course of the stepped transfer there is a substantial performance of the contract, then B, the intermediary party, is considered as having entered into a land transaction and a LTR is required. Therefore, the provisions can only apply where there is no substantial performance of the contract between A and B.

13.31 A lease for seven years or more can be self-certified provided there is no premium and no rent of any monetary value (s 77(2A) of FA 2003 as substituted by s 164 (1) of FA 2006). A peppercorn rent is of no monetary value. In practice, HMRC accepts that any rent of less than £1 is of no monetary value as the SDLT forms take no account of pence.

13.32 In para 13.23 above the circumstances of an actual lease following on from the triggering of an agreement for lease by substantial performance were considered. It is possible in these circumstances that there will be no chargeable consideration at the point of the actual grant of the lease due to the entire rent being taxed, at the point of substantial performance of the agreement for lease, perhaps triggered because of entry onto the land under licence. Where there is no premium, the impact of overlap relief, which acts by reducing the rent, may reduce the chargeable consideration to nil. In these circumstances, the actual lease can be self-certified but it is understood that HMRC expects a covering explanatory letter to be appended to the return to the Land Registry.

13.33 There are a number of circumstances where supplementary returns may be required, which are dealt with in Chapter 11 in more detail. An example is where the consideration was originally wholly or partly uncertain, unascertained, or contingent, but then becomes ascertainable. This may happen, for example, when a transaction is subject to a further payment, such as on the grant of a planning permission, or when an abnormal increase in rent becomes payable, or where a turnover rent falls to be finally quantified within the first five years of a rental period. There is now a general requirement to submit an additional return (SDLT 4) where the transaction involves either mixed or non-residential property. This requirement is detailed in the HMRC Guidance Note SDLT 6.

13.34 In these circumstances when the position becomes known, it might result in an increase in the tax payable. Alternatively, it might create an SDLT liability where none previously existed, for example, where cumulative rents payable from year-to-year have accumulated to cross one of the SDLT thresholds. In either case, the situation is the same as if either completion or substantial performance had occurred. The purchaser must, therefore, provide an LTR to HMRC within 30 days in exactly the way as set out in para 13.2 *et seq* above.

13.35 Any additional tax chargeable is calculated at whatever rate is in force at the further effective date and, as normal, must accompany the LTR.

13.36 Under s 80(4) of FA 2003, if the effect of the situation is to reduce the tax payable, by which is meant the tax paid with the original LTR, then the purchaser can request a repayment of tax, together with interest at the prevailing statutory rates from the initial date of payment.

13.37 In any circumstance where some form of disqualifying event triggers a clawback of tax because a relief has been withdrawn, then a further LTR is required, together with any tax payable. These types of situation would include withdrawal of reconstruction or acquisition relief, withdrawal of group relief, or the withdrawal of charities relief. In this case, the effective date would be the date on which HMRC served a notice withdrawing the relief. (Thereafter, the taxpayer would have 30 days to file the new LTR (s 81 of FA 2003)).

13.38 A similar situation occurs if a later linked transaction requires a purchaser to submit a further LTR in respect of the earlier transaction or transactions. The effective date is the date of the later transaction that brings about the need for the LTR (s 81A of FA 2003).

13.39 There are special regulations in respect of LTRs, or indeed any other documents relating to tax provided to HMRC, where that document is destroyed, lost, or is in some way made illegible. In these circumstances, the HMRC can treat the LTR as not having been properly delivered and may require a replacement LTR to be made (s 82 of FA 2003).

13.40 As this situation could potentially trigger penalties or a further demand for tax, if the purchaser can prove that the tax has already been paid (when the original defective document was filed), then relief against double taxation or, indeed, a repayment of overpaid tax can be achieved. Section 82(4) of FA 2003 requires the taxpayer to prove these issues to the satisfaction of the General or Special Commissioners. However, in practice, a reference to the General or Special Commissioners should not be necessary if the matter can be demonstrated to the satisfaction of the officer of the board concerned.

14 HMRC Powers and Compliance

A. Introduction

14.1 This chapter deals sequentially with Her Majesty's Revenue and Customs' (HMRC) powers of enquiry and its ability to enforce compliance with the statutory requirements, together with related penalties. The relevant legislation is found in Schedules 10, 11, 12, 13 and 14 to Finance Act 2003 (FA 2003) introduced by ss 78, 79, 91, 93 and 99. HMRC's own powers stem from the provisions of s 113 of FA 2003, these powers have been modified and to an extent extended by the provisions of s 82 *et seq* of Finance Act 2007 (FA 2007). Some matters such as appeals and those matters contained within Schedule 11 A are dealt with separately in Chapter 15.

B. Duty to keep and preserve records

14.2 Part 2 of Schedule 10 to FA 2003 (paras 9 to 12) requires a person responsible for delivery of an Land Transaction Return (LTR) to maintain those records needed to deliver a correct and complete return. Such records must be preserved for at least six years after the effective date of the transaction, or until any later date on which any enquiry into the return is complete or, if there is no enquiry, HMRC no longer has the power to enquire into the return (see section C below).

14.3 The records include documents relating to the transaction, namely any contract or conveyance, and any supporting maps, plans or similar documents, together with records of any relevant payments, receipts and financial arrangements. This would seem to include mortgage documentation.

14.4 It is not required that the actual documents should be kept, only the information contained therein. Information kept in copy form is admissible as evidence in any proceedings before the Commissioners.

14.5 The penalty for failure to keep and preserve records is a fine not exceeding £3,000. No penalty can be incurred if HMRC are satisfied that the facts that it reasonably required to be proved can be provided by other documentary evidence.

14.6 This would seem to indicate that, although there is a requirement to keep records, it is sufficient to be able to prove the facts of the individual transaction by copies or other evidence, provided there is sufficient information to establish the proof. This is particularly so in a case where HMRC seek to investigate or to challenge.

C. Enquiry into a return

14.7 Part 3 of Schedule 10 to FA 2003, paras 12 and 13, provides for HMRC to give a notice of enquiry to a purchaser if it wishes to enquire into a LTR submitted by the purchaser. Notice must be given before the end of the enquiry period. In some circumstances HMRC may issue a penalty notice prior to the commencement of an enquiry, this may be in circumstances where HMRC believes that the consideration declared on the SDLT 1 is less than it should otherwise have been. This and other areas are dealt with in Chapter 18 which deals with specific anti-avoidance legislation and the General Anti-Avoidance Regulations (GAAR).

14.8 The enquiry period is a period of nine months after the filing date if the return is delivered on time, or nine months after the date on which the return was delivered, if delivered late. Where an amendment has been made to an LTR, the period extends for nine months from that point.

14.9 An LTR, which is the subject of one notice of enquiry, may not be the subject of another. However, an amendment to an LTR allows the making of a further notice of enquiry in respect of that amendment. When an enquiry commences it continues until the point of the Completion of Enquiry, at which stage a closure notice is issued, pursuant to para 23 of Schedule 10 of FA 2003.

14.10 Some commentators take the view that, as stamp duty land tax (SDLT) is a transaction-based tax, HMRC does not have power to challenge the amount of consideration given in an arms length transaction. However, para 13 of FA 2003 states:

> An enquiry extends to anything contained in the return, or required to be contained in the return that relates (a) to the question of whether tax is chargeable in respect of the transaction, or (b) to the amount of tax so chargeable.

14.11 As the amount of tax payable relates to the consideration in the transaction, be it rent or purchase price, it would seem that HMRC has the power to enquire into the amount of consideration paid. If this were not the case, then challenging transactions where the consideration paid is close to, and just under, a band or relief level, would not be possible. On the other hand, if a vendor agrees to accept a payment at that level, it is unlikely that HMRC could charge SDLT on the higher sum that the vendor could have obtained. The parties would, of course, need to satisfy HMRC that there was no other concealed consideration.

14.12 The only exception is in the case of an amendment to a return, where the notice of enquiry is made at a point where it is no longer possible to enquire into the initial LTR or, after an enquiry into the LTR has been completed, the enquiry into the return is limited to those matters to which the amendment relates.

D. Notice to produce documents

14.13 Schedule 10 to FA 2003, paras 14 to 16, provides that, following a notice of enquiry into an LTR, HMRC may give notice in writing to the purchaser, requiring the purchaser to produce such documents in the purchaser's possession or power, and to provide HMRC with such information, and in such form, as HMRC may reasonably require for the purposes of the enquiry.

14.14 A notice to produce documents, which may be made at the same time as the notice of enquiry, must specify a time of not less than 30 days within which the purchaser is to comply with it. In practical terms, it is unlikely that, on launching an enquiry,

HMRC would not seek to obtain all relevant material to assist its enquiry. Delivery and service of documents follow normal HMRC procedures, as set out in s 84 of FA 2003. SDLT is a tax based on chargeable consideration and disputes are often based on a perception of value, or the component values inherent in the total consideration. Most value-based taxes are referred to the Valuation Office Agency (VOA) for discussion and negotiation, during the course of which supporting information and evidence is produced as part of the process. In SDLT cases HMRC often seek to serve notices under Schedule 10 prior to referring the case to the VOA, which does not aid the early settlement of the case.

14.15 The purchaser, in complying, may produce photocopies or facsimiles of the documents rather than originals. However, HMRC may, by further notice, require the originals to be produced for inspection. Similarly, the time-limit of not less than 30 days applies. In practical terms, this may create difficulties for the taxpayer as the notice will specify documents which HMRC believes, or assumes to exist, but which may not, in reality, exist. For example, calculations of the individual values of component parts of a transaction per para 4 of Schedule 4 to FA 2003 (see Chapter 10), which may have been the subject of a global calculation of the actual consideration paid. In these circumstances the taxpayer has no alternative but to appeal against the notice.

14.16 HMRC may copy, or take extracts, from any document produced to it. However, a notice under this paragraph does not require a purchaser to produce documents relating to any pending appeal by them, or referral to the Special Commissioners to which they are a party.

14.17 A purchaser has the right to appeal against a notice to produce documents (para 15 of Schedule 10). The notice must be in writing and made within 30 days of the issue of the notice appealed against. It must be made to the officer of the board by whom the notice was given. If the taxpayer does not appeal against the notice given under para 14 of Schedule 10, at the end of the 30-day appeal period, the taxpayer loses all right to make further appeals. Therefore, HMRC has an open ended right to demand whatever further information it requires, in whatever form it requires it to be delivered. This, as mentioned in para 14.15 above, may place the taxpayer in a difficult position as the taxpayer is required to produce documents which may not exist.

It is obviously unacceptable to require a taxpayer to produce or create documents which did not exist at the point of submitting the LTR.

14.18 An appeal against a notice to produce documents is determined in the same way as an appeal against assessment. The Commissioners may set aside the notice where the documents or information sought are shown not to be reasonably required for the enquiry. However, the Commissioners may confirm the notice where satisfied that it relates to documents or information reasonably required for the purposes of the enquiry. In this respect it is possible to have the information requested limited to such documentary information as actually exists and is relevant to the enquiry. Practical experience has demonstrated that once HMRC has made an assumption as to the information in a taxpayer's possession, it will not readily be convinced to the contrary.

14.19 A notice confirmed by the Commissioners must be complied with within 30 days from the date of determination. The decision of the Commissioners on this issue is final.

14.20 Failure to comply with a notice to produce documents may lead to a penalty of £50, and if the failure continues thereafter a daily penalty may apply, which cannot exceed £30 if the penalty is set by HMRC, or £150 if the penalty is set by the court. These penalties can only continue to the point at which the notice is complied with.

E. Amendments of LTRs during enquiry

14.21 Schedule 10 to FA 2003, paras 17 to 23, provides that HMRC may, during an enquiry into an LTR, decide that the amount stated in the self-assessment of tax is insufficient and that there may be a loss of tax unless matters are rectified. HMRC can give notice in writing to the purchaser to amend the assessment and to make good the deficiency. This would seem to be a likely course in most circumstances, as HMRC is unlikely to wish to enquire into a return unless it perceives that there is an underpayment of tax.

14.22 Where the enquiry relates to an amendment of an LTR, the notice can only apply to any under self-assessment which is a consequence of the amendment.

14.23 An enquiry is in progress from the date on which notice of the enquiry is given, until the day on which the enquiry is completed. Once the appropriate notices have been served within the requisite time periods, it would appear that the length of time over which the enquiry may continue is open ended.

14.24 During the course of an enquiry, a taxpayer has the right to amend the LTR themselves. The amendment, while it does not restrict the scope of the enquiry, may be taken into account as part of the enquiry.

14.25 If the amendment affects the amount of tax included with the LTR, it does not take effect while the enquiry is in progress. If HMRC state in the closure notice, that it has taken the amendment into account in formulating its notice, or if its conclusion is that the amendment is incorrect, then it shall not take effect. Otherwise, the amendment takes affect when the closure notice is issued.

14.26 Both HMRC and the taxpayer have the right to refer any questions concerning the subject of the enquiry to the Special Commissioners for determination. The taxpayer and HMRC must make a notice of referral jointly, in writing, to the Special Commissioners.

14.27 The notice of referral must specify the questions being referred. However, more than one notice may be given in relation to an enquiry, and referrals to the Special Commissioners may be made at any point during the enquiry period.

14.28 Both HMRC or the purchaser may withdraw a notice of referral. This must be done in writing, both to the other party to the referral and to the Special Commissioners, and must be prior to the hearing by the Special Commissioners of the issues in the referral notice. It would, therefore, appear that, once the hearing of the Special Commissioners has commenced, it must proceed to its conclusion, and neither of the parties to the appeal can seek to withdraw their notice during that time.

14.29 During a referral to the Special Commissioners, HMRC is unable to make a closure notice while proceedings continue. HMRC may not make an application for the Special Commissioners to direct that they should make such a notice. In this context, proceedings are in progress from the point when notice of referral is given and has not been withdrawn, and the questions referred to in the notice have not yet been determined.

14.30 When a question referred to in a notice is determined by the Special Commissioners there is no possibility of the determination being varied, set aside or subject to any appeal (this disregards any power or permission to appeal out of time).

14.31 The determination is binding on the parties to the referral. HMRC is, therefore, bound to take the determination into account in reaching its conclusions and in formulating amendments to the LTR under enquiry.

14.32 Once the question raised has been determined, it may not be challenged if there is a further appeal against the assessment raised by the HMRC on the closure of its enquiry. The only exception to this is if it had been determined as a preliminary issue in that appeal.

14.33 An enquiry is completed by the HMRC giving a closure notice telling the taxpayer that it has completed its enquiries and stating its conclusions. HMRC must state either that no amendment of an LTR is required, or state the amendments to the LTR required. The closure notice takes effect when it is issued.

14.34 As previously mentioned, an HMRC enquiry is open ended in time terms. However, a purchaser may apply to either the General or Special Commissioners for a direction to HMRC to give a closure notice within a specified period (para 24 of Schedule 10 to FA 2003). This type of application is dealt with in the same way as an appeal. The Commissioners shall direct HMRC to make a closure notice, unless they are satisfied there are reasonable grounds preventing HMRC giving a closure notice within the specified period.

F. HMRC determination if no LTR is delivered

14.35 Part 4 of Schedule 10 to FA 2003, paras 25 to 27, allows HMRC, if it believes that an LTR should have been but has not been made, to make a determination based on the information it has. The notice of determination must be served on the purchaser (or alleged purchaser), stating the date on which it is issued, the transaction to which it refers, and the amount of tax it believes is chargeable. No HMRC determination may be made more than six years after the effective date of the transaction.

14.36 The determination acts for enforcement purposes as if it were a self-assessment by the purchaser. This triggers the provisions of Schedule 10 to FA 2003, providing for tax related penalties, together with the provisions of s 87 of FA 2003, in relation to interest on unpaid tax, and the provisions of s 91 of and Schedule 12 to FA 2003, relating to the collection and the recovery of unpaid tax, interest and penalties. Penalties relating to the failure of the purchaser to deliver an LTR in the first place are treated as a separate issue.

14.37 The taxpayer is allowed, after the making of an HMRC determination, to deliver an LTR of their own in respect of the transaction. The self-assessment of tax included in and with the LTR supersedes the determination.

14.38 This is subject to two time limits, first where the LTR would be made more than six years after the day on which HMRC's powers of determination first became exercisable, or second, 12 months after the date of the notice of determination, whichever is the later.

14.39 If proceedings had begun for the recovery of tax under an HMRC determination, but before the proceedings are concluded the determination is superseded by a self-assessment, the proceedings can be continued as if it were seeking the recovery of so much of the tax payable under the self-assessment as has not been paid.

G. HMRC assessments

14.40 Part 5 of Schedule 10 to FA 2003, paras 28 to 32, allows HMRC, if it discovers a chargeable transaction where an amount of tax has not been assessed, or where the amount of tax is insufficient, or a relief has been given that is either wrong or has become excessive or disqualified, to make a discovery assessment in relation to the additional tax payable. This power is subject to restrictions dealt with in para 14.42 This power links with s 75A of Finance Act 2007 (FA 2007) inserted by the FA 2007, see also Chapter 16.

14.41 If HMRC discovers that there has been an excessive repayment of tax, or tax should not have been repaid, that amount can be recovered as if it were unpaid tax. This also applies to any interest paid to the taxpayer. These powers are also subject to some limitation as dealt with below. A general limitation applies where a mistake has been made in the completion of an LTR or amendment thereto, because the basis or method on which the tax liability had been computed was one generally in common use or practice at that time (para 30(5)).

14.42 Where an LTR has been delivered by a taxpayer, HMRC assessments, as detailed in paras 14.40 and 14.41 above, may only be pursued where the loss or excessive repayment is attributable to fraudulent or negligent conduct by the purchaser, a person acting on behalf of the purchaser, or a person who is a partner of the purchaser at the relevant time.

14.43 Additionally, assessments may be pursued where HMRC, at the time it makes the discovery, is not entitled to give a notice of enquiry into the LTR, or have completed its enquiries into the LTR, but could not have been reasonably expected, on the basis of the information available to it before that time, to have been aware of the true situation.

14.44 Information regarded as made available is either contained in the LTR made by the taxpayer, or is contained in any documents or information provided to HMRC during an enquiry, or it is information which could not reasonably be expected to be inferred by HMRC from information given in writing by the purchaser or the person acting on theirbehalf.

14.45 No HMRC assessment may be made if the situations set out above are attributable to mistakes in the LTR that had been made in accordance with the practice generally prevailing at that time.

14.46 Section 95 of FA 2003 (offence of fraudulent evasion of tax) and s 96 of FA 2003 (penalty for assisting in preparation of incorrect return, etc) define offences and penalties.

14.47 Section 95 provides that a person commits an offence if they are knowingly concerned in the fraudulent evasion of tax by themselves or another person. A person guilty of such an offence is liable to imprisonment for a term not exceeding six months, or a fine not exceeding the statutory maximum, or both. On conviction on indictment they are liable to imprisonment for a term not exceeding seven years, or a fine, or both. These fines would be in addition to any tax-related penalties it was thought proper to impose.

14.48 Section 96 provides that a person who assists in, or induces the preparation or delivery of information contained in an LTR, which they know will, or would likely to be used for any purpose of tax, and which they know to be incorrect, is liable to a penalty not exceeding £3,000. This section is drawn wide enough to include the professional advisors to the taxpayer, and would also seem to be capable of including the vendor of a property. For example, suppose a vendor and the purchaser agree that the value of a residential property is, for example, £495,000, and agree as a linked transaction that fixtures and fittings included in the sale have a separate value of, say, £25,000. Where it can be proved that these fixtures and fittings did not have such a value and the proper consideration should have been, say, £515,000, then if this agreement was made for the purposes of avoidance of tax, it would appear that both the purchaser would be liable to prosecution for an offence under s 95, and the vendor, together with both vendor's and purchaser's advisors who were aware of, or party to, this agreement, could be liable under s 96.

14.49 The general rule, following that for other taxation, is that no assessment may be made more than six years after the effective date of the transaction to which it relates. However, where the case involves fraud or negligence on the part of the purchaser, a person acting on their behalf, or a person who was their partner at the relevant date, then an assessment may be made up to 21 years after the effective date of the transaction to which it relates.

14.50 An assessment to recover excessive repayment of tax is not out of time where, during the course of a notifiable enquiry, it is made before the date the enquiry is completed, or in any case it is made within one year after the repayment in question was made.

14.51 Where the purchaser has died, any assessment on their personal representatives must be made within three years of their death. Furthermore, in this respect, the six-year rule applies where the transaction had an effective date of six years prior to the taxpayer's death.

14.52 Where an objection is to be made to the making of an assessment on the grounds that the time limits have expired, such an objection can only be made on appeal against the assessment itself.

14.53 Any notice of assessment must be served on the taxpayer. It must also state the tax due, the date on which the notice is issued, and the time within which any appeal against the assessment must be made.

14.54 Once a notice of assessment has been served, the assessment may not be altered except where expressly provided, as set out above.

H. Self-certification

14.55 Schedule 11 to FA 2003, paras 1 to 17, contains provisions for regulations covering self-certification which mirror those for the delivery of an LTR. There is a duty to keep and preserve records for the same period of six years, together with a duty to preserve information. Where there is a failure to keep and preserve records then the penalties are exactly as those set out in para 14.5 in regard to the self-assessment of tax. This is to prevent the maker of a self-certificate not being able to provide such evidence as HMRC may require to pursue an enquiry.

14.56 The rules concerning enquiries carried out by HMRC are exactly the same as those into an LTR. A notice of enquiry has to be served in the same way and has to detail the scope of the enquiry. Similarly, notices to produce documents for the purposes of the enquiry may be made and, as in the case of the LTR, the taxpayer may appeal against such notices.

14.57 Penalties for failure to produce documents are exactly the same as in the case of an LTR, and it is also possible that questions may be referred to the Special Commissioners during an enquiry. There are exactly the same requirements in the case of withdrawals of notices of referral, and referral enquiries have the same effects as set out for LTRs in paras 14.27 to 14.30.

14.58 There is the same procedure in relation to the determination by the Special Commissioners, and the taxpayer has the same rights in relation to closure notices as in the case of an LTR (para 14.33). There is also a provision to allow a taxpayer to apply to the General or Special Commissioners for a direction for HMRC to complete an enquiry and, as in the case of an LTR, HMRC may only resist such an application on the grounds that it has reasonable grounds for not giving a closure notice within a specified period (para 12).

I. Collection and recovery of tax

14.59 Schedule 12 to FA 2003, paras 1 to 3, deals with collection and recovery of tax.

14.60 A collector of taxes may demand tax which is due and payable, for which, if requested, the collector shall give a receipt.

14.61 In England, Wales and Northern Ireland, unpaid tax may be recovered by distraint and, in Scotland, by diligence. The procedures are essentially the same in that a warrant has to be issued either by a justice of the peace or, in Scotland, by the sheriff. The difference between Scotland and the rest of the United Kingdom is that diligence is an attachment, or an earnings arrestment, or an arrestment and action of furthcoming or sale. The position in the rest of the United Kingdom is that goods and chattels may be seized and sold by public auction in an attempt to satisfy the debt.

14.62 In England, Wales and Northern Ireland where the amount concerned does not exceed £2,000, it may be recovered summarily as a civil debt in proceedings brought by the collector of taxes.

14.63 In all circumstances, the proceedings are brought in the name of a collector of taxes, who has a duty to provide the correct documentation and certification to the court, following which the tax may be sued for and recovered from the person charged, either as a debt due to the Crown, or by any other means available, by proceedings in the High Court or, in Scotland, the Court of Session sitting as the Court of Exchequer.

J. Powers to call for documents or information

14.64 Schedule 13 to FA 2003 deals in detail with the mechanisms under which an authorised officer of the board may seek the delivery of documents or information in relation to an enquiry into an LTR. As explained in para 14.13 *et seq*, above, HMRC can issue a notice requiring the delivery of documents within a period of at least 30 days. If the taxpayer does not produce the information voluntarily, and HMRC decides that a formal notice requiring delivery of the documents or information is required, it needs the prior consent of a General or Special Commissioner. HMRC view the application of Schedule 13 powers as more onerous than Schedule 10 powers and only those officers with specific authorisation can trigger the use of those powers.

14.65 When making an application to the Commissioner, HMRC must provide the taxpayer with a written summary of the reasons for making the application. The Commissioner has to be satisfied that HMRC has reasonable grounds for believing the document or information will assist the collection of tax. HMRC may make copies or extracts from the documents or information obtained under such a notice (Schedule 13 to FA 2003, paras 1 to 5).

14.66 Provided the consent of the Commissioner is obtained, HMRC can also require a person to provide, or make available to it, documents that, in its reasonable opinion, are relevant to the tax liability of some other person. This might apply in the case of an executor where the taxpayer who is the subject to the enquiry is dead, or to an officer of a company which has been wound up. HMRC, in this situation, must be able to specify the actual documents which it requires and, again, must give at least 30 days for their production. Except in exceptional circumstances, HMRC must also inform the person concerned of the name of the taxpayer into whose affairs the enquiry is being made.

14.67 When seeking documents from a third person, HMRC has no power to seek information. Therefore, under these powers, HMRC cannot seek opinions or hearsay evidence from the person concerned. A copy of the third party notice, together with the reasons given to the Commissioner, must also be given to the taxpayer concerned although, when the information is sought from an executor or beneficiary, this would seem to be unnecessary.

14.68 Part 3 of Schedule 13 to FA 2003 gives HMRC powers to call for papers of a tax accountant. However, this power only exists where the tax accountant has been convicted in the United Kingdom for an offence in relation to tax, or has had a penalty imposed under s 96 of FA 2003 which relates to a penalty for assisting in the preparation of an incorrect return.

14.69 In order to access the tax accountant's papers, HMRC require the consent of a circuit judge. However, in these circumstances it can require access to documents containing information relevant to a tax liability of any of the accountant's clients.

14.70 Part 4 of Schedule 13 to FA 2003 sets out significant restrictions on HMRC's powers for obtaining information. These limitations are briefly:

a. personal records or journalistic material

b. information in relation to an impending appeal

c. information held by a barrister, advocate or solicitor (unless the Board of HMRC itself issues the notice)

d. deliveries of original documents as copies are sufficient (however, an officer of the board can insist on viewing the originals)

e. documents originating more than six years before the date of notice (in fraud cases the commissioners may direct otherwise)

f. documents subject to legal privilege

g. documents belonging to an auditor or tax advisor which contain information or advice given to the client for the purposes of tax advice. There is an exception to this, if these documents contain items which explain information already delivered to HMRC either on a LTR or under a notice.

14.71 It is interesting to speculate as to how the legislation contained in the Finance Act 2004, in relation to the registering of tax avoidance schemes with HMRC, interacts with this part of Schedule 13.

14.72 Part 5 of Schedule 13 to FA 2003 gives special powers to the Board of HMRC itself to give a notice requiring the delivery of documents or provision of information where it has reasonable grounds for believing that there may be serious prejudice to the proper assessment or collection of tax. In this respect, it does not require the prior consent of the Commissioners.

14.73 Part 6 of Schedule 13 provides that, in certain circumstances where there is reasonable ground for suspecting an offence involving serious fraud in connection with SDLT has or is about to be committed, HMRC can apply to a circuit judge to order documents or information to be provided within 10 working days, or a shorter or longer period, as provided in the order. In Scotland, the application is to a sheriff and, in Northern Ireland, to a county court judge. The taxpayer must be given notice of the application as they are allowed to appear before the judge or sheriff to oppose it.

14.74 Linked to these powers is the ability of HMRC, under Part 7 of Schedule 13, to apply to the judge or sheriff for a search warrant to look for evidence that a serious fraud in relation to tax has, or is about to be, committed. Interestingly, under s 107 of FA 2003, a search warrant cannot be issued in relation to premises occupied for the purposes of the Crown.

14.75 Items subject to legal privilege, however, are not allowed to be seized, unless those items are held for the intention of furthering a criminal purpose (Schedule 13 to FA 2003, para 48).

14.76 The need to obtain a search warrant in order to seek documents or information should not be confused with HMRC's power to inspect any property for the purpose of ascertaining its market value. Section 94 of FA 2003 gives this power to an officer of the board. However, it states that the person having custody or possession of the property should permit the authorised officer to inspect it at such reasonable times as the board may consider necessary. This means that the officer of the board should make a proper appointment, with prior notice, before inspecting the property, and cannot expect to be given access to the property merely by turning up on site. However, it should be remembered that anyone who delays or obstructs an officer of the board is committing an offence, and may be subject to a fine not exceeding level 1 on the standard scale.

14.77 It is an offence, under Part 8 of Schedule 13, to falsify, conceal, destroy or otherwise dispose of a document that has been required by a notice of enquiry. This also includes causing or committing a third person to carry out these offences. As described in para 14.20, this may give rise to an initial fine, followed by an application by the Commissioners of a daily penalty, although, in the case of a document being destroyed, it is hard to see how the taxpayer could comply with such a notice.

14.78 Where a person, who may be either the taxpayer or a third person, fraudulently or negligently delivers, or provides, or makes available an incorrect document or information, they are liable to a penalty not exceeding £3,000 (s 93 of FA 2003).

15 Appeals, Relief for Overpayment, Claims not Included in LTRs

A. Introduction

15.1 This chapter deals with the mechanism of appeals in relation to stamp duty land tax (SDLT) cases where relief is sought because of excessive assessment, and the procedure where a claim is made for deferment of tax in any situation where the claim cannot be made either in, or by, a subsequent amendment to an Land Transaction Return (LTR).

B. Appeals and other proceedings

15.2 The statutory provisions are contained in Schedule 17 to Finance Act 2003 (FA 2003), as amended by the SDLT (Appeals) Regulations 2004 (SI 2004/1363), which both interlink with Part 1 of the Taxes Management Act 1970 to align the appeals procedure for SDLT with the generality of the taxes acts. Schedule 17 to FA 2003, para 2, gives the bodies that may be appealed to as the General Commissioners, the Special Commissioners, the General or Special Commissioners, or the Lands Tribunal.

15.3 In this context the Lands Tribunal means the Lands Tribunal for England and Wales, the Lands Tribunal for Scotland, and the Lands Tribunal for Northern Ireland.

15.4 Some cases lie to the General or Special Commissioners and some to the Special Commissioners. There is, of course, a significant difference between the composition of a body of General Commissioners and a Special Commissioner. In the case of General Commissioners, these are normally three lay persons appointed in a particular area or region and who are advised on matters of law by a legally qualified clerk.

15.5 A Special Commissioner normally sits alone, although may have an adjudicator sitting with them. The Special Commissioner will be qualified either as a solicitor or a barrister and will have a specialised knowledge of taxation law.

15.6 Generally speaking, all appeals have to be made within an appropriate time limit, and by the giving of a notice of appeal in writing to the officer of the board who either gave a notice of assessment or issued a closure notice. This appeal, in common with others throughout SDLT legislation, has to be made within 30 days of the date on which the notice, assessment or amendment was issued, rather than received.

15.7 Under certain circumstances, an appeal may be made out of time. This can be done with the consent of the officer of the board concerned, who must be satisfied that there was reasonable excuse for not bringing the appeal within the time limit. Furthermore, they must be satisfied that the application for consent to bring proceedings out of time was made without unreasonable delay.

15.8 Should the officer of the board not accept an appeal made out of time, the matter may be referred for determination by the Commissioners. The application is generally made to the General Commissioners, even if there is a right to elect that the Special Commissioners shall hear the appeal itself.

15.9 Appeals can be brought against:

a. an amendment or a self-assessment made under Schedule 10 to FA 2003, para 17, which is where Her Majesty's Revenue and Customs (HMRC) seek, during an enquiry, to amend an LTR immediately to prevent loss of tax (see para 14.21). In this case, the Commissioners do not hear the appeal until the HMRC enquiry is complete and a closure notice issued

b. a closure notice made by HMRC, where either HMRC has stated the conclusion that it has reached, or stated the amendment which it wishes to be made to the LTR

c. a discovery assessment.

15.10 Section 115 of FA 2003, together with Schedule 17, para 2, allow the Lord Chancellor to make regulations concerning the matters to be referred to which body of Commissioners, as appropriate, or to the Lands Tribunal. The Lord Chancellor may also make regulations under Schedule 17 to FA 2003, para 6, concerning both the composition of the General or Special Commissioners and also allowing the Commissioners certain powers.

15.11 The Regulations, as mentioned above, are the SDLT (Appeals) Regulations 2004 (SI 2004/1363) which define the nature of the appeals to be directed to the various commissioners or tribunals.

15.12 The Commissioners may join, as a party to the proceedings, a person who would not otherwise be a party (an advisor or partner of the taxpayer for example). They may require a party to the proceedings to provide information and make documents available for inspection by themselves, any party to the proceedings, or an officer of the board.

15.13 They also may require persons to attend a hearing to give evidence and produce documents generally in relation to the proceedings, and which allow the Commissioners to review their decision.

15.14 They are also enabled to impose penalties, to make orders for the determination of recovery of such penalties, and also to hear appeals against penalties. For example; the making of a daily penalty by HMRC (see para 13.13) requires an application to have been made to, and granted by the Commissioners.

15.15 In general terms more minor cases, probably involving procedural issues, liability to make an LTR or the like, would be referred to the General Commissioners. More complex matters, where the regulations specifically allow, or, where the parties to the appeal agree, may be referred to the Special Commissioners.

15.16 Matters of valuation pertinent to an appeal would be referred by the Commissioners to the Lands Tribunal. In these cases where the Lands Tribunal makes a determination in relation to value, the matter is passed back to the Commissioners to be incorporated in their decision.

15.17 In some cases, the decision of the Commissioners is final. However, Schedule 17 of FA 2003, para 9, allows the Lord Chancellor to make regulations about the finality of decisions in certain circumstances. These set out the way in which a determination or award by the Commissioners may be questioned.

15.18 Generally speaking, there is a provision for an appeal to lie to a higher court on any question of law arising from a decision of the Commissioners. Normally, the decision of the Commissioners would not be challengeable in a higher court if it relates to question of fact on which they have ruled.

15.19 The exception to this stems from the case of *Edwards (Inspector of Taxes)* v *Bairstow and Harrison* (1956) 36 TC 207. The principle established here was that, generally, the courts would not disturb the finding of fact made by either a body of commissioners or a tribunal, unless the conclusion was so strange that no reasonable tribunal could have arrived at it. This concept is in fact hard to substantiate unless it can be demonstrated that the tribunal concerned have ignored, or misinterpreted, evidence of fact fundamental to the decision reached.

15.20 The Lord Chancellor also has the power to make regulation concerning costs. Costs are not generally awarded before the General Commissioners, each party to the appeal bearing their own costs. Before the Special Commissioners, where legal representation and the calling of expert witnesses is the norm, then the Commissioners may award costs, as they see fit. Before the Lands Tribunal, the member may make an award as to costs and, in all cases where costs cannot be agreed between the parties, they may be directed to be taxed on the high court scale.

C. Appeals against HMRC decisions on tax

15.21 Part 7 of Schedule 10 to FA 2003 deals with the making of appeals against HMRC decisions on tax. Appeals may be made against:

a. amendments of self-assessment to prevent loss of tax (see para 14.21)

b. a conclusion stated or an amendment made by a closure notice
c. a discovery assessment
d. an assessment to recover excessive repayment (see para 14.41).

As set out above, these appeals lie to either the General or Special Commissioners, and the Lands Tribunal.

15.22 Where an appeal has been made it may be settled by agreement between the taxpayer, the taxpayer's advisors and HMRC. There are basically three forms of agreement.

- That the decision appealed against should be upheld without variation.
- That the decision appealed against should be varied in a particular manner.
- That the decision appealed against should be discharged or cancelled.

15.23 In each case, matters shall be treated as if, at the time the agreement had been made, the Commissioners had determined the appeal in the manner that the agreement provides for (Part 7, Schedule 10 to FA 2003, para 37).

15.24 Following the making of an agreement, the appellant has 30 days to notify HMRC in writing if, on reconsideration, they wish to withdraw from the agreement.

15.25 Where the initial agreement is not in writing, it cannot be acted upon until the fact that an agreement has been come to, and the terms agreed, are confirmed by notice in writing given either by HMRC to the appellant, or by the appellant to HMRC. Once the agreement has been confirmed in writing, the 30-day period mentioned above then commences.

15.26 If the appellant notifies HMRC orally or in writing that they do not wish to proceed with the appeal, HMRC have 30 days to give the appellant notice, in writing, stating that itis unwilling that the appeal shall be withdrawn. If he does not do so, the matter shall be treated as if an agreement had been reached that the decision under appeal shall be upheld without variation.

15.27 Part 7, Schedule 10 to FA 2003, para 37(5), allows agreements and notifications by the appellant to include those that may be made on their behalf by their advisors, or those acting on their behalf.

15.28 Where an appeal is made, the tax chargeable under the assessment or amendment remains due as if there had been no appeal. However, if the appellant believes that they are overcharged to tax by the decision against which they are appealing, the appellant may apply to the Commissioners for a postponement of payment pending the determination of the appeal (see also para 15.31).

15.29 The notice must be given to the relevant officer within 30 days of the specified date. It must state the amount of the overcharged tax and the grounds on which the taxpayer believes this to be true.

15.30 An application for deferment may be made outside the 30-day limit if circumstances change, and those changes lead to the appellant believing they are being overcharged.

15.31 The amount of tax that is deferred is any amount in excess of the amount the taxpayer accepts as being payable. Where, for example, the taxpayer believes that the tax payable is £1,000, and HMRC believe the tax payable is £1,500 (which forms the basis of appeal), deferment would relate to the £500 overpayment. Therefore the taxpayer should, if they have not already done so in returning the original LTR, pay the amount of tax accepted as outstanding.

15.32 An exception to this would be in the case of a dispute as to whether a self-certification certificate should be appropriate, in which case the whole of the amount of tax payable could possibly be deferred. This would also seem to be the situation where HMRC sought to withdraw a relief in relation to a claim made by the taxpayer.

15.33 There are a number of variations to the above situation. For example, the Commissioners may rule that some tax is to be paid, and their determination on this point then becomes the date on which that tax should be paid.

15.34 When a determination is finally made by the Commissioners, and that leads to HMRC issuing a notice of the tax payable following the determination, then, unless the matter is subject to further appeal, the tax becomes payable on that date.

15.35 It is open to the taxpayer, or their advisor, and HMRC to agree the amount of tax that should be properly postponed, in which case the agreement is treated as if the Commissioners had made the agreement, which must be in writing (Part 7, Schedule 10, to FA 2003, para 40).

D. Relief in case of excessive assessment

15.36 Part 6, Schedule 10 to FA 2003 makes provision for a taxpayer, who believes they have been assessed for tax more than once in respect of the same transaction, to make a claim for relief against any double charge.

15.37 A taxpayer, who believes they have paid excessive tax by reason of a mistake in an LTR, may make a claim to HMRC for repayment.

15.38 The claim must be made within six years of the effective date of the transaction. However, no relief is available if it relates to a mistake as to the basis on which the LTR ought to have been computed, particularly if this computation was made on the basis of or in accordance with practice generally prevailing at the time when it was made.

15.39 Similarly, no relief is available if there was a mistake in a claim or election included in the return.

15.40 The basic concept is that, if the way in which a particular computation is made alters due to changes in approach, common practice, or the law over a period of time, no retrospective claim may be made by attempting to apply what has then become current practice to the historic LTR.

15.41 HMRC must have regard to all circumstances to the case, particularly whether the granting of relief would result in amounts being excluded from a charge to tax. This is so where a change in law practice would provide a different result from that current when the original LTR was made.

15.42 Any appeal following the decision of HMRC lies to the Special Commissioners who shall determine the claim in relation to the principles set out in Part 6 of Schedule 10 to FA 2003. These are:

a. to ensure the claim is made within six years

b. to ensure the claim is not against the effect of using a method of computation that was in general use at the time of the making of the original LTR
c. to consider if the granting of relief would result in amounts being excluded from tax
d. to have regard to all the circumstances of the case.

15.43 If proceedings have begun for recovery of tax under an HMRC determination but, before the proceedings are concluded, the determination is superseded by a self-assessment, the proceedings can be continued as if they were seeking the recovery of so much of the tax under the self-assessment as has not been paid (Schedule 10 to FA 2003, para 27(3)).

E. Claims not included in returns

15.44 It is possible that a claim for overpaid tax may come about in a situation where an amendment may not be included in an LTR, or may be made by a supplementary LTR or amendment.

15.45 Normally, such a claim would relate to contingent or unascertainable consideration, where a future event would lead to a further payment or increase in consideration to an unquantified level. However, where the amount is quantified, or may be ascertained from the face of the contract, SDLT on the whole is payable with the original LTR, and is not capable of deferment (see also Chapter 12).

15.46 An example of this would be where a rent was set at £100,000 per annum for the initial term, with the proviso that, when a specific event occurred, it would rise to £200,000 per annum, for example, the pedestrianisation of the locality, or the opening of a new car park. SDLT payable would be determined by allowing for this eventuality. If the specified event is either cancelled or indefinitely postponed, the rent will not therefore rise, as predicted, and the taxpayer has overpaid tax. Schedule 11A to FA 2003 allows for the making of claims in such circumstances.

15.47 A claim must be made in an appropriate form and allow for a declaration that the details provided are, to the claimant's information and belief, correct. It must also:

a. state the amount of tax to be repaid

b. provide enough information for HMRC to determine the validity of the claim
c. provide documentary evidence as may be required.

15.48 The claimant must keep, from the original effective date of the transaction concerned, proper records to substantiate the claim, and preserve them until the later of:

a. 12 months from the date of claim
b. where there is an enquiry, until the enquiry is complete (this includes enquiries into amendments to a claim)
c. where the claim is amended when there is no longer an HMRC right to enquire into the amendment.

15.49 The same penalty of up to £3,000 applies for non-compliance as it does to LTRs.

15.50 The procedure follows, thereafter, almost exactly the same form as the making of an LTR with provision for:

a. amendments by taxpayer and HMRC
b. correction of a claim by HMRC
c. giving effect to claims and amendments
d. notice of enquiry
e. notice of documents for the purposes of an enquiry
f. appeals against notices
g. penalties for failure to comply with notices to produce documents, the penalty being £50 with a subsequent daily penalty of £30, (no permission of the Commissioners being required to levy the daily penalty in this case)
h. making a closure notice on completion of an enquiry
i. seeking a direction from the Commissioners for the enquiry to be completed
j. the affect of an amendment and appeals against amendments being the same.

15.51 Any appeal against a conclusion stated or amendment made lies to the Special Commissioners. This is subject to Schedule 10 to FA 2003 concerning relief not being allowed.

16 Connected Persons and Companies

16.1 Initially, the main purpose of the connected company's legislation, in relation to stamp duty land tax (SDLT), was to prevent the setting up of special purpose vehicle companies by arrangement between connected parties, in an attempt to avoid, to mitigate, or to minimise the amount of SDLT properly payable. However, this now has an impact where connected persons are involved in partnerships, see Chapter 9, and prior to the Budget 2007 where connected persons entered into exchanges of property, see Chapter 4 at para 4.29.

16.2 Section 53 of Finance Act 2003 (FA 2003) concerns "Deemed market value where transaction involves connected company". Where the purchaser is a company and the vendor is connected to the purchaser, or some or all of the consideration for the transaction consists of the issue or transfer of shares in a company with which the vendor is connected, the chargeable consideration for the transaction shall be taken to be not less than the market value of the subject matter of the transaction as at the effective date of the transaction.

16.3 Hence, where the purchaser is a company, and either the vendor is connected with that company, or part or all of the consideration for the transaction is by way of shares in a company with which the vendor is connected, the value transferred is assessed under the deemed market value provisions. This means that the actual value of the asset will be taken as the consideration rather than a notional value ascribed to the shares. The valuation approach that will be adopted by Her Majesty's Revenue and Customs (HMRC) is detailed in Chapter 10.

16.4 Exemptions from the deemed market value rule are contained in s 54 of FA 2003. These cases are where the purchaser company holds the subject property as a trustee or as a manager of trusts; the purchaser company is connected by way of the terms of a settlement; the transaction is, or is part of, distribution of the assets of the vendor company, whether or not in connection with its winding up, and where the subject property has not been involved in a transaction where group relief was claimed by the vendor company in the previous three years.

16.5 For the purposes of this section, the meaning of connected persons is as set out in ss 839 and 840 of the Income and Corporation Taxes Act 1988, which are quoted in detail in Appendix 3.

16.6 The main forms of connection are by "blood" or by "business", including direct blood relations or relations created by marriage. In relation to business connection, this can be either by way of trusts and settlements or by way of shared business relationships or partnerships. Much of the anti-avoidance legislation in both SDLT and in general taxation is aimed at preventing the exploitation of such relationships to avoid, mitigate or minimise taxation.

16.7 For the purposes of SDLT, many transactions which are between connected persons, in the general run of events, are exempt from SDLT. These include gifts, testamentary dispositions, dispositions under intestacy, dispositions of property to beneficiaries under a trust in accordance with the terms of that trust, certain transactions on the ending of a marriage, and certain transactions following a person's death.

16.8 Although not mentioned in the FA 2003, which pre-dates the legislation, the meaning of connected persons, the legislation in this area has been somewhat extended by s 993, etc of the Income Tax Act 2007 and s 179 of the Income Tax (Trading and Other Income) Act 2005. In the case of s 179, the extension relates mainly to corporate bodies, in the case of ss 993, 994 and 995, it brings together both the provisions of s 839 and other provisions relating to partnerships, trustees and companies.

17 Right to Buy Transactions, Shared Ownership Leases, Alternative Property Finance, etc

A. Introduction

17.1 Section 70 of Finance Act 2003 (FA 2003) introduces Schedule 9 to FA 2003, which deals with right to buy transactions, shared ownership leases, and the Scottish equivalent of rent to mortgage and rent to loan transactions. They are described in Chapter 1.

17.2 Section 71 of FA 2003 deals with transactions involving registered social landlords (RSLs), while ss 72 and 73 of FA 2003 deal with alternative property financing. Sections 74 and 75 deal with collective leasehold enfranchisement and community right to buy schemes.

17.3 All of the above are subject to various forms of relief and, in some cases, variations to the basis on which chargeable consideration is assessed. Many of the transactions (with the exception of alternative property financing) may fall below either the residential threshold of £125,000 or, more particularly, the higher threshold of £150,000 in designated disadvantaged areas (DDAs). As each category has its own specific rules, they are included herein for the sake of completeness. The main thrust of the legislation is to ensure that tax is paid on the actual purchase price paid rather than market value, and ignores the concept of contingent consideration specifically using Schedule 9 to FA 2003 to disapply the provisions of s 51.

B. Right to buy transactions

17.4 A right to buy transaction comprises the sale of a dwelling at a discount, or the granting of a lease at a discount, by a relevant public sector body (Schedule 9 to FA 2003, para 1).

17.5 Relevant bodies are:

a. a Minister of the Crown, Scottish Ministers, Northern Ireland Departments

b. local housing authorities within the Housing Act 1985: English county councils: Scottish councils within s 2 of the Local Government etc (Scotland) Act 1994; district councils within the Local Government Act (Northern Ireland) 1972

c. the Housing Corporation: Scottish Homes: The Northern Ireland Housing Executive: a RSL: a housing action trust established under Part 3 of the Housing Act 1988

d. the Commission for the New Towns: a development corporation established under the New Towns Act 1981 or the New Towns (Scotland) Act 1968; an urban development corporation under s 135 of the Local Government Planning and Land Act 1980; a new town commission under s 7 of the New Towns Act (Northern Ireland) 1965: the Welsh Development Agency

e. a police authority within s 101(1) of the Police Act 1996 or s 2(1) or s 19(9)(b) of the Police (Scotland) Act 1967: the Northern Ireland Policing Board

f. an Education and Libraries Board in Northern Ireland, the UK Atomic Energy Authority, any person mentioned in paras 9, (k), (l) or (n) of s 61(11) of the Housing (Scotland) Act 1987: and any body prescribed by Treasury order.

17.6 A right to buy transaction also arises where the sale or grant of a lease of a dwelling is in pursuance of a preserved right to buy, where the vendor is a private sector landlord against whom a right to buy is exercisable under the provision of Part 5 of the Housing Act 1985 in England and Wales, and under s 61 of the Housing (Scotland) Act 1987.

17.7 The purchaser must be the qualifying person and the dwelling the qualifying dwelling-house in relation to the preserved right to buy.

17.8 In a right to buy case, s 51 of FA 2003 does not apply (contingent consideration), and chargeable consideration does not include any consideration payable if a contingency may or has occurred.

17.9 Therefore, for right to buy properties, the chargeable consideration is limited to the initial purchase price only. Where an early disposal of the property leads to an actual or potential claw back of the discount given by the vendor, this is ignored for stamp duty land tax (SDLT) purposes, and no further rent to loan is required.

17.10 In a right to buy purchase, where the vendor is a RSL, any purchase grants in respect of the disposal by the RSL at a discount are not included as chargeable consideration.

C. Shared ownership leases

17.11 A shared ownership lease is not defined but, in brief, is one that satisfies the provisions of Schedule 9 to FA 2003, para 2, which sets out what type of lease is a shared ownership lease. The conditions are:

a. the lease must be of a dwelling
b. the lease must give the lessee exclusivity of possession or use
c. the lease must provide for the lessee to acquire the reversionary interest
d. the lease must be granted partly for a rent and partly a premium.

17.12 The premium must be calculated by reference to the market value of the dwelling or is a sum calculated by reference to that value. Paragraph 2(5) provides that s 118 of FA 2003 (meaning of market value) does not apply in relation to the references in this paragraph to the market value of the dwelling. This means that the market value of the property is always taken as its market value with vacant possession, and the element of share value for the premium is taken as a mathematical fraction thereof. The more complex approach, set out in Chapter 10, does not therefore apply (see paras 10.12 to 10.42).

17.13 The lease must also include a statement of market value, together with the sum calculated by reference to it and by reference to which the premium is calculated. Paragraph 3 to Schedule 9 of FA 2003 provides that where a valid election has been made and any SDLT has been paid, any subsequent transfer of the reversion under the lease agreement is exempt from charge.

17.14 If the lease passes the above tests there are two further conditions. First, the lease must be granted by a qualifying body. These are a local housing authority, a housing association, a housing action trust, the Northern Ireland Housing Executive, the Commission for New Towns or a development corporation.

17.15 Alternatively, the first condition may be met if the lease is granted in pursuance of a preserved right to buy (see paras 17.6 and 17.7 above).

17.16 Second, the lessee must elect for SDLT to be charged in accordance with para 2 (see para 17.12). This election must be made in the Land Transaction Return (LTR) filed in respect of the lease (or in an amendment to that return). Once made, the election is irrevocable and cannot be altered by a subsequent amendment to the LTR (para 2(3)). SDLT will be payable on the premium. Once paid, any subsequent transfer of the reversion to the lessee will be exempt from SDLT (para 3). This is fair since the lessee paid tax on the sum derived from the market value with assumed vacant possession at the outset.

D. Staircasing transactions

17.17 Like shared ownership leases, staircasing is not specifically defined in the legislation but, like shared ownership leases, each lease has to satisfy various conditions. These are broadly similar to those for shared ownership leases, and are set out in Schedule 9 to FA 2003, para 4.

- The lease must be of a dwelling.
- The lease must give exclusivity of possession or use.
- The lease must allow the lessee to have the lease altered on payment of a sum, so that the rent payable is reduced (staircasing).
- The lease must be granted partly for a rent and partly for a premium.

17.18 The premium is calculated by reference to a premium payable for a lease granted on the same terms on the open market, substituting the minimum rent for the rent payable under the lease, or a sum calculated by reference to that premium. The minimum rent is the lowest rent that could become payable if the maximum sum is paid.

17.19 The lease must state the minimum rent and the premium obtainable on the open market or the sum calculated by reference to that premium.

17.20 Where the lease meets the above tests there are two further conditions. First, the lease must be granted by a qualifying body (see para 17.14 above). Alternatively, the first condition may be met if the lease is granted in pursuance of a preserved right to buy (see paras 17.6 and 17.7 above).

17.21 Second, the lessee must elect for SDLT to be charged in accordance with para 4 in the same way as a shared ownership lease (see para 17.16 above). SDLT will be payable on the premium obtainable on the open market assuming the minimum rent is payable. Subsequent staircase payments will be exempt up to 80% of the market value of the dwelling, as defined in para 17.12 above.

E. Rent to mortgage or rent to loan transactions

17.22 The terms are defined in Schedule 9 to FA 2003, para 6(2) and (4). The manner in which chargeable consideration is determined is provided by para 6(5).

17.23 A rent to mortgage scheme arises from Part 5 of the Housing Act 1985 (HA 1985) . Instead of buying the property for a capital sum, the tenant is deemed to have borrowed that sum as a mortgage. Thereafter, the payment of rent is treated as payment of capital so that the payments repay the "mortgage" by instalments. Hence, a rent to mortgage transaction involves the transfer of a dwelling, or it can be the grant of a lease of a dwelling to a person exercising a right to acquire on rent to mortgage terms under Part 5 of HA 1985.

17.24 The chargeable consideration is the price that would be payable on the transfer or on the price payable for the grant of a lease, calculated in accordance with s 126 of HA1985, the right to buy scheme as it applies to rent to mortgage transactions. The price is discounted market value.

17.25 Rent to loan transactions apply to Scotland and are similar to the rent to mortgage scheme. They involve the execution of a heritable disposition in favour of a person exercising the right to purchase a house under the rent to loan scheme under Part 3 of the Housing (Scotland) Act 1987. The chargeable consideration is the price that would be payable calculated in accordance with s 62 thereof for the outright balance.

F. Relief for certain purchases by registered social landlords

17.26 Where a RSL purchases property from a qualifying vendor or with the assistance of public subsidy, the purchase is exempt from SDLT (s 71 of FA 2003). The RSL must be controlled by its tenants. This is one where the majority of the board members are tenants of properties owned or managed by it. A board member means a director of a company, a member of a body corporate, a trustee of a body of trustees, or, if otherwise, a member of the committee of management.

17.27 When the transaction is funded with the assistance of a public subsidy, wholly or partly, this can be by way of any grant or other financial assistance.

17.28 The public subsidy must originate from one of the following:

a. s 25 of the National Lottery etc. Act 1993 (application of money by distributing bodies)
b. s 18 of the Housing Act 1996 (social housing grants)
c. s 126 of the Housing Grants, Construction and Regeneration Act 1996 (financial assistance for regeneration and development)
d. s 2 of the Housing (Scotland) Act 1988 (general functions of the Scottish Ministers)
e. article 33 of the Housing (Northern Ireland) Order 1992.

17.29 A qualifying vendor must be a qualifying body, which is one of the following as contained in s 71(3) of FA 2003.

- Other RSLs.
- Housing action trusts established under Part 3 of the Housing Act 1988.
- A principal council under the Local Government Act 1972.
- The Common Council of the City of London.
- The Scottish Ministers.
- A council under s 2 of the Local Government etc (Scotland) Act 1994.
- Scottish Homes.
- The Department for Social Development in Northern Ireland.
- The Northern Ireland Housing Executive.

Since qualifying bodies enjoy public funding, a transaction with one of them is akin to a transaction funded with the assistance of public subsidy.

G. Alternative property finance

17.30 Exemption from multiple charges to SDLT are available for two types of alternative property finance (ss 72 and 73 of FA 2003). They apply principally to Islamic mortgages. The intention is that SDLT will impact in the same way as if the property had been acquired under a conventional mortgage arrangement. The scope of the legislation has been extended and amended by Finance Acts 2005–2008. This is partly to increase its scope to cover various variations used in practice and partly to prevent this part of the legislation being used for avoidance purposes.

17.31 The exemptions are available to individuals using one of three types of alternative property finance, one of which has two variants. Exemptions must be claimed in relation to each transaction, using an LTR, and showing the chargeable consideration as nil. Broadly speaking, the transactions involve the calculation of a grossed up sum which includes the finance costs and the property is held on a "leasehold" basis with regular payments being made across the term of the arrangement. The legislation is contained in ss 71A, 72, 72A and 73 of FA 2003, legislation is introduced in FA 2008 to prevent an abuse of the system using a subsidiary company owned by the financial institution to transfer the property without payment of SDLT, or payment at a reduced rate, to the benefit of a third party purchaser.

i. Land sold to a financial institution and leased by prior arrangement to an individual

17.32 An individual enters into an arrangement with a financial institution (see para 17.42 below), the financial institution purchases a major interest in a property for, say, £100,000, then grants a lease or sublease to the individual for 25 years, subject to an option exercisable at the discretion of the individual to have the reversion transferred.

17.33 The terms of the lease are that, over the 25-year term, some £225,000 is paid to the financial institution. At the end of the lease, having paid the agreed sum, the reversion is transferred to the individual. Effectively, the £225,000 is equivalent to both interest and principal in the context of a 25-year loan. The first acquisition of the property may be subject to SDLT in the normal manner. However, where a financial institution acquires the property from the individual with whom it is entering into the arrangement or from another financial institution which acquired the interest as part of an alternative property finance scheme, then it, too, may claim SDLT exemption under s 72(2) of FA 2003.

17.34 Basically, what has happened is that the first transaction is the acquisition of a major interest in land by a financial institution, the second transaction is the grant of a leasehold interest to the individual, and the third transaction is the transfer of the initial major interest back to the individual. The transactions are consequential upon each other, and the third is a condition of the leasing arrangement. Detailed rules are set out in s 72(1) to s 72(4) of FA 2003.

17.35 There are specific provisions to ensure that the relief is only available to an individual or individuals (see paras 17.37 and 17.43 below).

17.36 Exemption from SDLT is potentially claimable on each transaction and, while institutions owning the asset may transfer the property between themselves without disturbing the individual's rights to exemption, such a transaction may attract SDLT in the normal manner. An individual may, however, transfer the alternative property finance from one institution to another institution by arrangement, without disturbing the chain of reliefs. In such a case the acquiring institution may claim exemption.

17.37 Where any part of the chain of transactions is broken, or the rules about it being for the use of individuals as an alternative form of finance, in substitution for a conventional mortgage arrangement are breached, then relief may be denied. For example, there is no exemption if the individual holds the lease or sublease as a trustee, and any beneficiary is not an individual, or the individual holds the interest with a partner who is not an individual.

ii. Land sold to a financial institution and re-sold to an individual

17.38 An individual enters into an arrangement with a financial institution, the institution buys a major interest in a property for £100,000 and sells it back to the individual for £225,000, and the individual pays off the amount in regular instalments over 25 years in accordance with a legal mortgage granted by the individual to the institution.

17.39 Exemption is potentially available on the first and second transactions, provided the rules set out in s 73 of FA 2003 are followed. Effectively, these are the same as for land sold to a financial institution and leased to an individual. However, the initial acquisition of the property, either by the institution or individual, may be subject to SDLT in the normal way.

17.40 To obtain the exemptions, the first transaction must be the purchase of a major interest in land by the institution from the individual, or from another financial institution which purchased the interest as part of an alternative property finance scheme. The second transaction is the sale of that interest back to the individual, the individual granting the financial institution a legal mortgage over the asset.

17.41 Provided the relationship between the individual and the institution is retained, the exemptions continue to be available. However, if the second transaction is a disposal or transfer to a third party, relief from SDLT is not available.

iii. Definitions

17.42 Financial institutions are:

a. banks within the meaning of s 840A of the Income and Corporation Taxes Act 1988

b. building societies within the meaning of the Building Societies Act 1986
c. a wholly owned subsidiary of (a) or (b) above.

17.43 Individuals are living persons. Groups of individuals may claim the exemption such as partnerships. However no company, apart from the financial institution, may be party to the transaction. Beneficiaries of a deceased individual may step into the deceased's shoes.

17.44 In Scotland, references in ss 72 and 73 of FA 2003 to freeholds are taken as being "the interest of the owner" (or a *dominium utile* until abolished), and to leaseholds are "a tenant's right over or interest in a property subject to a lease".

17.45 Legal mortgage, in England and Wales, is as defined in s 205(1)(xvi) of the Law of Property Act 1925, in Scotland as a standard security, and, in Northern Ireland, "a mortgage by conveyance of a legal estate or by demise or sub-demise or a charge by way of legal mortgage".

iv. Interest held by a financial institution

17.46 The Finance Act 2007 (FA 2007) introduced s 73B, the interest held by a financial institution as a result of the "first transaction [F1] per s 71A (1)(a), s 72 (1)(a) or s 72A is an exempt interest within the meaning of schedule 3". The effect of this is that dealings in that interest, or an interest derived from that interest are not notifiable.

17.47 The interest ceases to be an exempt interest if:

- the lease or agreement mentioned in s 71C (1)(c), s 72 (1)(b) or s 72 (1)(b) ceases to have effect
- the right under s 71A (1)(d), s 72 (1)(c) ceases to have effect or becomes subject to a restriction. This is the right to require a transfer of the financial institution's interest to the person entering into the arrangements.

17.48 The F1 interest is not an exempt interest if group relief was claimed on the first transaction, neither is the F1 interest an exempt interest in respect of:

- the FI itself, so that even if the FI is exempt from charge it remains notifiable
- a further transaction or third transaction within the meaning of s 71A (4), s 72(4) or s 72A(4), which will be generally exempt from charge but remain notifiable.

17.49 Section 73B has effect in relation to anything that would, but for the exemption, be a land transaction with an effective date on or after 22 March 2007. It is immaterial when the F1 interest originally came into existence.

17.50 The above is taken from HMRC's Stamp Duty Land Tax Manual, the main thrust here is that the operation of alternative property finance does not sit comfortably with SDLT in general and the subsequent variations to the legislation are an attempt to ensure that the rules are not manipulated to achieve avoidance of tax properly payable, while allowing the reliefs to remain for genuine transactions.

H. Collective enfranchisement by leaseholders

17.51 Relief for collective enfranchisement is available, for SDLT purposes, when collective enfranchisement is carried out by a group of leaseholders acting together as a company known as a right to enfranchise company (s 74 of FA 2003). The right of collective enfranchisement is derived from the Landlord and Tenant Act 1987 and the Leasehold Reform, Housing and Urban Development Act 1993, and relates to flats.

17.52 There is a specific method laid down for calculating the amount to be taken as chargeable consideration and the consequential SDLT payable. This is for the total consideration to be apportioned equally to the total number of flats. This will commonly result in a lower amount of SDLT to be payable.

I. Crofting community right to buy

17.53 Where two or more crofts are being acquired under the right to buy provisions of Part 3 of the Land Reform (Scotland) Act 2003, the relevant consideration is divided by the number of crofts being acquired which determines the rate of tax and the tax chargeable (s 75 of FA 2003). As for collective enfranchisement, this will commonly reduce the amount of SDLT payable.

J. Commonhold

17.54 Commonhold is a way of owning freehold properties which have communal facilities such as blocks of flats. Commonhold was created by the Commonhold and Leasehold Reform Act 2002 which was enacted on 27 September 2004. It was found that the practical application of commonhold did not sit well with SDLT as enacted.

17.55 The registration of common parts in the name of a commonhold association and the registration of a unit in the name of a unit holder (see para 17.50) are all land transactions, being the acquisition of chargeable interests by operation of law. Generally there is no chargeable consideration for these transactions, or is negligible.

17.56 Commonhold land can be registered either by unit holders, who are generally the individuals who own each flat in a block of flats and the freehold interest therein. Or it can be registered by a developer who is common proprietor until the property is sold and the interest registered in the name of the commonhold association. It is considered by Her Majesty's Revenue and Customs (HMRC) that where the transferee is a company then the interest will have consideration deemed to be at its market value as per s 53 of FA 2003.

17.57 This is at variance to the way in which HMRC views a commonhold association because it is not able, as a matter of law, to deal in or profit from the common parts but must maintain and insure them.

18 Stamp Duty Land Tax and Trusts

A. Introduction

18.1 Section 105 of Finance Act 2003 (FA 2003) introduces Schedule 16 to FA 2003, which deals with trusts and the responsibilities of trustees. First, this chapter sets out Her Majesty's Revenue and Customs' (HMRC's) view on and interpretation of the legislation. In this area of the legislation a number of commentators have suggested that HMRC's interpretation is flawed both in law and practice. At the end of this chapter the areas of concern are highlighted.

18.2 Gifts, testamentary dispositions, dispositions under the intestacy law, dispositions of property under a trust in accordance with the terms of that trust, and most acquisitions by operation of law are not chargeable to stamp duty land tax (SDLT) (para 1 to Schedule 3 of FA 2003).

18.3 There is a land transaction when land passes to a beneficiary under a will, or by virtue of the law on intestacy. In England, Wales and Northern Ireland this transaction is effected by means of an assent or an appropriation. In Scotland, a docquet or docket is normally used but there may be a disposition that is a deed transferring the property. Paragraph 3A to Schedule 3 of FA 2003 provides that the acquisition of property by a person:

- in or towards satisfaction of their entitlement under or in relation to the will of a deceased person is exempt from SDLT

- on the intestacy of a deceased person is exempt from SDLT.

18.4 This exemption does not apply to the extent that the beneficiary gives consideration other than the assumption of secured debt. Secured debt is debt that, immediately after the death, was secured on the land, for example a mortgage, which was not paid off on death. The exemption applies whether the transfer is to a sole beneficiary or to joint beneficiaries.

i. Trusts and powers

18.5 Schedule 16 of FA 2003 sets out the responsibilities of trustees and how SDLT applies in relation to interests in trust and to the acquisition of a chargeable interest to the exercise of power of appointment or discretion. The responsibilities of the trustee and beneficiaries in relation to SDLT depend upon the kind of trust in which the land is held. For the purposes of SDLT there are two basic types of trust defined in para 1 to Schedule 16. These are:

- bare trusts which include nominee arrangements
- settlements.

ii. Bare trusts

18.6 A bare trust in England, Wales and Northern Ireland or simple trust in Scotland is one in which each beneficiary is absolutely entitled against the trustees of the property comprised in the trust. The phrase "absolutely entitled" broadly means that:

- the beneficiary may acquire or receive the trust property immediately by giving the requisite notice to the trustees in accordance with the terms of the trusts

- the trustees have no power over or right to deal with the trust property without the permission of the beneficiary who enjoys absolute entitlement where:
  - the trustees are required to meet certain outgoings and expenses of the trust
  - the trustees refused to make trust property available to the beneficiary until such outgoings or expenses have been met.

The determination or whether the beneficiary is absolutely entitled to the trust property as against the trustee is not effective. Two or more persons may be absolutely entitled as against the trustees to trust property, providing that each of them has the rights to the trust of the property described above. Where a person acquires a chargeable interest as bare trustee, SDLT applies as if the trust was invested in, and the acts of the trustee in relation to it, where the acts of the person or persons for whom they are trustee.

18.7 An example of the nature of a bare trust is set out in the HMRC, *Stamp Duty Land Tax Manual.*

- Nominees Ltd, a corporate trustee, holds a house on trust for X who is a beneficiary.
- The house is let.
- X is entitled to the rents which are X's income.
- X is entitled to do whatever they wish with the house.
- Nominees Ltd cannot do anything with the house other than that they are instructed to do by X.
- Nominees Ltd hold the title to the house and the Land Registry shows the house as held in the name of Nominees Ltd.

Therefore, if a bare trustee or nominee acquires property on behalf of the beneficiary that would be treated as an acquisition by the beneficiary, the beneficiary being the person liable for SDLT and responsible for submitting the Land Transaction Return (LTR) to HMRC.

iii. Settlements

18.8 A settlement for SDLT is any trust arranged other than a bare trust. There are many types of trust arrangements, common examples are interest in possession trusts. This type of trust exists when a beneficiary, known as an income beneficiary, has a right to the income of the trust as it arises. The trustees must pass all of the income received, less any trustee's expenses and tax to the beneficiary. A beneficiary who is entitled to the income of the trust life is known as a life tenant in England, Wales and Northern Ireland or a life-renter in Scotland. The income beneficiary does not have any rights over the capital of this kind of trust; normally the capital will pass to a different beneficiary or beneficiaries at a specific time in the future or following a specific future event.

B. Discretionary trusts

18.9 Trustees who are discretionary trusts generally have discretion about how to use the capital and income of the trust. They may be required to use any income for the benefit of the particular beneficiaries. They can decide how much they get paid and to which beneficiaries. The beneficiary of a discretionary trust has no right over or interest in the capital or income of the trust.

i. Other trusts

There are other kinds of trusts that are not bare trusts and so will be regarded (by HMRC) as settlements for the purposes of SDLT. These include accumulation of maintenance trusts, mixed trusts, which are the mixtures of more than one type of trust and trusts set up under the laws from jurisdictions. When the trustees of a settlement acquire land, the trustees were regarded as the purchases for SDLT, therefore all the normal rules regarding notification and payment relate to responsible trustees.

18.10 The responsible trustees in relation to a land transaction are any persons who are trustees at the effective date of the transaction and any persons who subsequently becomes a trustee. No penalty or interest on such a penalty would be recovered from a person who did not become a responsible trustee until after the relevant time. The relevant time is, in relation to a daily penalty or interest on that penalty, the beginning of that day and in relation to any other penalty or interest on that penalty, the time when the act of emission that caused the penalty to be payable occurred.

18.11 Where payment for power of appointment or discretion to be exercised is made, it will be treated as consideration for the acquisition of a land or interest through the exercise of a power or discretion. Therefore, where consideration is provided to the trustees in exchange for the exercise of their power of appointment, such as an interest in land passes out of the trust to a person, the consideration so provided is treated as consideration for the acquisition for the relevant interest in land.

ii. Nil-rate discretionary trusts

18.12 A nil-rate band discretionary trust is commonly established under a will, a typical form is a legacy, not exceeding the nil-rate band for inheritance tax (IHT) to be held by trustees on a discretionary trust for a specified class of beneficiary. Where the pecuniary legacy is discharged by the payment of the specified sum, no SDLT issues arise. If the legacy is satisfied by the transfer of property, say the matrimonial home, then an SDLT liability may arise. The transfer of land to a residuary beneficiary under a will is a land transaction for SDLT purposes.

18.13 The main question is whether any chargeable consideration is given. Some examples shown in the *Stamp Duty Land Tax Manual* are set out below.

- The trustees accept the surviving spouse's or civil partner's promise to pay in satisfaction of the pecuniary legacy and in consideration transfer land to the survivor. The promise to pay is consideration for SDLT purposes.
- A personal representative, acting for the survivor makes a promise, as above, on their behalf. The promise to pay is still consideration for SDLT purposes.
- Land is transferred to the survivor and they charge the property with payment of the amount of the pecuniary legacy. The consideration is the amount promised, up to the market value of the property.
- Land is transferred to the survivor, the property is charged as above, and the trustees accept this charge in satisfaction of the legacy. The charge is monies worth and therefore consideration.
- The personal representatives charge the land with the payment of the pecuniary legacy. They agree with the trustees that there is no right to enforce payment against the owner of the land for the time being. The trustees accept the charge in satisfaction of the legacy, and the land is transferred to the survivor subject to the charge. There is no charge to SDLT provided there is no change in the rights or liabilities of any person in relation to the debt secured by the charge.

C. Relevant trustees

18.14 Any land transaction or return or self-certification to the Land Registry in relation to a land transaction may be made or given by any one or more of the trustees who are the responsible trustees in relation to the transaction. The trustees making the return or self-certification are known as the relevant trustees. All relevant trustees must make the declaration on return or self-certificate confirming it is completed correct.

D. Appeal and enquiry powers

18.15 The appeal and enquiry powers require that any notice, assessment or amendment to be given by HMRC must be given to each of the relevant trustees. However, an appeal or closure application may be made by any of the relevant trustees. See also Chapters 13 and 14.

E. Changes in the composition of trustees for continuing settlement

18.16 For SDLT purposes, HMRC treat trustees of a settlement as a single and continuing body of persons, in the same manner as they are treated for capital gains tax.

18.17 It therefore follows that for continuing settlement the change in the composition of trustees is not a land transaction. This means, in particular, that there is no charge on such an occasion where a trust property is secured by a mortgage or other borrowing as there is no land transaction for SDLT purposes on a change in the composition of trustees for continuing trusts, and a land transaction term should not be completed.

18.18 Where a change should result in an application to the Land Registry, or the keepers of the registers of Scotland, a covering letter should accompany the application, unless it is obvious that the documents relate solely to such a change.

F. Transfers between pension funds

18.19 The following paragraphs deal with the situation where there is a transfer of assets and obligations from one pension fund to another, for example, on the payment of a statutory cash equivalent transfer value for an individual, or on a merger of funds. The transfer of land from the trustees of one pension fund to the trustees of another is the acquisition of a chargeable interest under s 48 of FA 2003. This means that it is within the scope of SDLT. The normal charge to SDLT arises on the consideration given for the land transaction.

18.20 There are no special rules for pension funds. SDLT will only be due where there is chargeable consideration for the transaction. In the view of HMRC, the assumption by the transferring fund or by the trustees of the transferring fund of obligations to provide benefits is not chargeable consideration.

18.21 If the transferring fund or trustees of the transferee fund give other considerations, in the form of money or monies worth, then that will be chargeable consideration. It would also be chargeable consideration if the transfer or obligations were in consideration to a defined monetary sum to be satisfied by the release of obligations by the former trustees.

18.22 In respect of borrowing and mortgages, a pension fund may borrow money and grant a mortgage and other charge over land as security. For SDLT purposes, it is necessary to consider the borrowing and the mortgage separately, then the context of a transfer as described above.

18.23 Therefore with regard to borrowing, if the transferee fund or trustees of the transferee funds assume an existing liability of the transferor fund or the trustees of the transferor fund repay borrowing or otherwise bring about the release of the transferor fund or trustees of the transferor fund from the debt and they do so as part and parcel of such a transfer, then in these circumstances, HMRC will not treat para 4 to Schedule 8 of FA 2003 as meaning that there is a chargeable consideration given for the land transaction.

18.24 Mortgages and other legal charges are security interests and dealing with them, including their creation and release, are specifically exempt from SDLT.

18.25 Although a land transaction for no consideration is exempt from notification under s 77 of FA 2003, it must be self-certified in accordance with s 79(3) (b) of FA 2003 if registration of the title is necessary.

18.26 Where pension fund trustees acquire land, other than as part of a transfer described in paras 18.13 to 18.21 above, where or not from another pension fund, SDLT is due on the consideration given in the normal way.

G. Consideration for exercise of power of appointment or discretion

18.27 Para 7 to Schedule 16 of FA 2003 applies where a chargeable interest is acquired.

- The exercise of a power of appointment.
- The exercise of discretion vested in trustees of a settlement.

This provides that any consideration given by the person in whose favour the power or discretion is exercised is treated as consideration to the acquisition.

18.28 The above provision has no application to normal trust transactions; it is intended however, to deal with the unusual case where a person pays trustees, or someone else, in order that the power or discretion may be exercised in their favour.

H. Persons acting in a representative capacity

18.29 Section 106 of FA 2003 provides for persons acting in a representative capacity to form SDLT functions, including making returns and dealing with correspondence. Additionally, a person holding a power of attorney can sign returns on behalf of the purchaser by virtue of s 81 B of FA 2003 (see also Chapter 1). The trustee, guardian or other person having the direction, management or control of the property of an incapacitated person is responsible for discharging any obligations in relation to a transaction affecting that property, to which the incapacitated person would be subject if they were not so incapacitated. Also, the person having control may retain, out of money they hold on behalf of the incapacitated person, sums to meet any payment they are liable to make and so far as they are not reimbursed is entitled to be indemnified in respect of any such payment. The term "incapacitated" is not defined in the SDLT legislation but in HMRC's view, the definition in s 108 of the Taxes Management Act 1970 (TMA 1970), which also applies in relation to s 72 of TMA 1970, on which s 106 of FA 2003 is based, should be followed.

18.30 The parent of a guardian or minor is responsible for charging any obligations of the minor that are not discharged by the minor themselves.

18.31 The personal representative of the person who is a purchaser under a land transaction is responsible for discharging the obligation for the purchaser in relation to the transaction. They may deduct any payment made by them in respect of SDLT out of the assets and effects of the deceased person.

18.32 A receiver appointed by a UK court and therefore having direction and control of any property is responsible for discharging any obligations in relation to any transaction affecting that property, as if the property were not under the direction and control of the court.

I. Relief on transfer of land to a unit trust

18.33 It should be noted that in the 2006 Budget, there was the abolition of the relief under s 64(a) of FA 2003 for transfers of a property to a unit trust on its creation. This had effect from the 22 March 2006 and was intended to prevent the use of offshore property unit trusts as a vehicle to enable purchasers to enquire interests in UK land free of SDLT.

18.34 This legislation did not seek to change the transfer of units in a unit trust, which own UK land and bring it within the charge to SDLT.

J. Alternative views

18.35 As stated above, an exchange of land between two family trusts with or without payment of any equality money or the acquisition of land for chargeable consideration by one trust to another will attract SDLT in the ordinary way. In particular, with an exchange, each side of the transaction is chargeable as per s 47 of FA 2003.

18.36 Several commentators have identified potential problems in the situation where within a single settlement there is an exchange or acquisition of land as between distinct funds or sub funds. This problem arises because para 1(1) to Schedule 16 of FA 2003 is widely drawn.

18.37 It is understood that the HMRC view, in a life interest case, was that a transfer of land from one sub fund to another sub fund within the same settlement was a land transaction for SDLT purposes. Furthermore, in this situation the purchaser is not the trustee but the beneficiary, the consideration being valued according to the actuarial value of the relevant life interest. However, seepara 18.41 below.

18.38 Many commentators felt that this view cannot be correct, except perhaps in the rare example of a settled Land Act Trust, because in reality the beneficiaries would not be party to the transaction. The situation can also vary depending on whether or not the trustees of the sub funds are the same body or persons or separate individuals.

18.39 The question therefore is — who is the purchaser? A person cannot be a purchaser unless they have given consideration for, or are party to, a transaction (s 43(5) of FA 2003). Therefore, it is hard to understand how, in what was apparently the HMRC view, a beneficiary could be the purchaser as they would have played no part in the appropriation, having no powers to appropriate and their consent not being required, and the beneficiary will not have taken any steps to carry out the appropriation.

18.40 As previously stated, dispositions of properties under a trust (see para 18.2 *et seq*) and most acquisitions by operation of law are not chargeable to SDLT. It would therefore seem that the purchaser, for SDLT purposes, can only be a trustee and, if in the case of sub funds, trustees of each fund are the same individuals, it is hard to see how there can be a purchaser. However, if each one has separate trustees it might be arguable that the definition of settlement of the trust that is not a bare trust (para 1(1) to Schedule 16 of FA 2003) is wide enough to embrace sub funds. Therefore, if each sub fund is treated as a separate settlement then the acquiring trustees potentially are purchasers under s 43(5) of FA 2003, the case then becomes a notifiable transaction with a requirement to deliver an LTR and pay tax under s 76 of FA 2003. This is on the assumption that the current HMRC view of the application of para 1(1) to Schedule 16 is correct.

18.41 In the technical news of September 2006, the previously understood HMRC position was redefined. HMRC stated:

> It is possible for a settlement to be made up of sub-funds, each sub-fund having particular beneficiaries. Concern has been expressed [see above] that the reallocation of property between sub-funds, which trustees may undertake from time to time to balance the interests of

beneficiaries, may give rise to an SDLT charge. We are satisfied that there is no SDLT charge where the beneficiaries are purely passive and play no part in the reallocation. FA 2006 [section 162 of FA 2006 inserted para 8 to schedule 16 of FA 2003] also makes it clear that a mere requirement that the beneficiaries should consent to the reallocation does not of itself give rise to an SDLT charge. HMRC will apply this treatment to all reallocations since the introduction of SDLT.

18.42 The above highlights one of the fundamental problems in relation to SDLT. It is a tax that has been subject to continuous variation and alteration by way of regulations and HMRC's evolving interpretation since its introduction in 2003. However, in the case of SDLT, there is little or no guidance to be gained from decided cases in relation to the legislation as it stands. This leads to an unhelpful situation as far as taxpayers are concerned under the self-assessment regime and particularly in respect of compliance with the requirements of that legislation. However, in the above situation HMRC has moved significantly closer towards the views expressed by practitioners.

19 The Disclosure Regime and Other Anti-Avoidance Measures

A. Introduction

19.1 Stamp duty land tax (SDLT) legislation itself is based on anti-avoidance. Prior to the introduction of SDLT in December 2003 it was generally considered that stamp duty was a tax capable of being avoided. This was basically because it was a tax based on the stamping of documents rather than a tax based on transactions.

19.2 The procedure adopted with SDLT has been to introduce change based on regulation in the form of statutory instruments. This is then followed up in the subsequent Finance Act by primary legislation. As a practical note, regulations only have a "life-span" of 18 months then they must be incorporated into primary legislation. Anti-avoidance generally has been a driving force in taxation legislation over the last decade. This has led to most taxes; such as income tax, corporation tax, and VAT being subject to the introduction of general anti-avoidance legislation commonly referred to as GAAR. The basic legislation being in Part 7 to the Finance Act 2004 (FA 2004).

19.3 Since the introduction of SDLT in 2003, every successive Budget has dealt with changes, tightening the legislation and regulations in order to counter various forms of avoidance and perceived avoidance. This chapter deals with, first, the ruling (advice) which can be sought, both pre-transaction and post-transaction, from Her Majesty's Revenue and Customs (HMRC), together with the changes brought in sequentially by the Finance Acts 2004, 2005, 2006 and 2007.

B. Code of Practice 10

19.4 Code of Practice 10 (COP 10) details how HMRC offers information and advice to taxpayers to help them understand obligations, so that they can get their tax affairs correct and pay tax on time. One of the situations in which HMRC provides advice is on the interpretation of legislation passed in the last five Finance Acts. HMRC believe this is where COP 10 will be of most value to those dealing with SDLT.

19.5 The full COP 10 booklet is available on the HMRC website at *www.hmrc.gov.uk/pdfs/cop10.htm* with hard copies available from the Stamp Office. Taxpayers wishing to use COP 10 information or advice would be well advised to read this leaflet. Among other things, the leaflet details information that needs to accompany a request for advice under COP 10. To place reliance on advice received from HMRC, the taxpayer has a responsibility to make full and proper disclosure. If the information provided is inaccurate, misleading or insufficient, then it is open to HMRC to vary the advice given, withdraw any clearances or, if necessary, seek to make a discovery assessment. This area has been subject to significant litigation and the leading case in relation to inaccurate and misleading information is *R* v *Commissioners of the Inland Revenue, ex parte Matrix Securities Ltd* [1993] EGCS 187. The leading case on incomplete disclosure is the Court of Appeal case of *Langham (Inspector of Taxes)* v *Veltema* [2004] EWCA Civ 193. This case is of particular interest as it involved the return of a property value (for income tax purposes) at what was subsequently discovered after initial acceptance to be at an undervalue.

19.6 The conclusions in the case of *Langham* v *Veltema*, as expounded in the Court of Appeal, created an inherent difficulty for taxpayers in complying with the law. Subsequently, HMRC issued guidance in relation to income and corporation taxpayers in a statement of practice in December 2004. While this guidance does not specifically mention SDLT, HMRC consider that it is also applicable to it. This illustrates a particular difficulty with value based taxes as an opinion of value, even if honestly held, may be at significant variance to that arrived at by the Valuation Office Agency (VOA) under the statutory basis (see also Chapter 10).

19.7 It should be appreciated that unlike stamp duty, HMRC does not have an adjudication function in relation to SDLT. This is why advice is (sometimes but not always) proffered rather than a ruling. The COP 10 booklet makes it clear that HMRC will not help with tax planning, or advise on transactions designed to avoid or reduce the tax charge that might otherwise be expected to arise.

19.8 Rulings given under COP 10 apply only to the specific transaction in question and do not relate to other transactions, which the taxpayers might regard as similar or identical. Requests for guidance under COP 10 should be sent to the customer service team at the Birmingham Stamp Office and should be addressed to the COP 10 Section. See Chapter 12 for the Birmingham Stamp Office address.

19.9 COP 10 rulings will, in many cases, automatically trigger an enquiry into the Land Transaction Return (LTR) by HRMC. This should be borne in mind when seeking an interpretation of either the legislation or regulations. Often, where advice is sought on a contentious point, HMRC will decline to advise on the grounds that the tax regime is based on self-assessment and it is for the taxpayer to decide.

C. Anti-avoidance 2004

19.10 The Finance Act 2004 introduced a new direct tax disclosure regime requiring promoters and, in some cases, taxpayers of certain tax planning arrangements to notify the arrangements or proposed arrangements to HMRC. The basic legislation is, now, in Part 7 to the Finance Act 2007 (FA 2007). The initial regime applied to schemes that sought to give income tax, corporation tax or capital gains tax advantages. It was then announced that this disclosure rule would be extended and adapted to apply to schemes and arrangements intended to avoid SDLT on commercial property transactions. Initially they were intended to apply to schemes made available or implemented after July 2005, the commencement date was in fact 1 August 2005.

D. Anti-avoidance 2005

19.11 The disclosure rules require promoters of tax avoidance schemes, such as accountants and lawyers, to provide HMRC with information about schemes and arrangements which might be expected to provide, as a main benefit of the scheme, an SDLT advantage in relation to a property that:

- is not a residential property within the definition of s 16 of the FA 2004
- has a market value of at least £5 million.

19.12 In contrast to the disclosure regime introduced by FA 2004 for SDLT purposes, HMRC do not issue a reference number and a promoter has no obligation to provide a reference number to the client taxpayer. This reduced obligation still provides HMRC with information about schemes but was intended to reduce the compliance burden on taxpayers. A drawback from the HMRC perspective is that this system does not normally identify the actual taxpayer who has used the scheme.

19.13 Taxpayers did not have an obligation to provide HMRC with information unless the following circumstances apply.

- The promoter is offshore.
- The users define the scheme as being in-house.
- The promoter is a lawyer who cannot make a full disclosure without revealing legally privileged material.

19.14 In relation to the last bullet point, the taxpayer can choose to waive privilege and request the lawyer concerned to make a disclosure if the taxpayer so prefers.

19.15 Following the extension of the discloser scheme to SDLT, HMRC appeared slow to act on disclosures. The only identifiable response was withdrawal of "Seeding Relief" in March 2006. This relief exempted the first transfer of property into a new unit trust scheme in consideration of an issue of units. What happened in practice was that new regulations and subsequent legislation have been introduced in order to make disclosed schemes ineffective. Schemes blocked with effect from 17 March 2005 are set out below.

19.16 Acquisition relief (Part 3, Schedule 7 of FA 2003). Changes were made dealing with the claw-back of group relief where the transferee company ceased to be a member of the same group.

19.17 Grants and leases from bare trustees to their principal (para 3 to Schedule 16 of FA 2003). The change has meant that the fact that one of the parties to the lease was a bare trustee would be ignored when determining the charge on the grant of a new lease.

19.18 A variation of a lease where the tenant gives charge to consideration falls to be treated as the acquisition of the chargeable interest of the tenant, not merely a variation to reduce the rent.

19.19 The use of repayable loans or deposits as consideration was formerly used as an avoidance advice. The contingency of any repayment, following the Finance Act 2005, was to be ignored.

19.20 Schemes seeking to reduce market value by a use of an encumbrance. This has an impact on sale and leaseback transactions. The changed rule required that the market value of the freehold or head lease was determined ignoring any effect of the leaseback or of any prior agreement to leaseback.

19.21 The use of subsale relief in alternative finance transactions (ss 45 and 73 of FA 2003). Subsale relief would no longer be available under s 45(3) of FA 2003 where the sale to the eventual purchaser was a second transaction that benefited from relief under s 73 to FA 2003. The rules in relation to alternative finance transactions have been significantly changed.

19.22 Partnership regulations were changed in circumstances where land was transferred into partnership and the transferor takes money out of the partnership within three years. This is comparable to the three-year rule, used in group relief for companies and the like.

19.23 The above were all given application from Budget Day 2005 in relation to land transactions and at the same time transitional provisions were published, which are no longer of particular concern. The main statutory instrument being SI 2005/1868 (SDLT Avoidance Schemes (Prescribed Descriptions of Arrangements) Regulations 2005).

E. Anti-avoidance 2006

19.24 The Finance Act 2006 (FA 2006) contained relatively little direct variation to SDLT in terms of anti-avoidance. There were two sets of specific measures; one already mentioned above was the withdrawal of seeding relief from Budget Day 2006. The second area was clarification in relation to SDLT in the area of partnerships and in particular the transfer of land to a partnership and the transfer of an interest in a partnership.

19.25 Changes were brought about in the FA 2006 in relation to property investment partnerships and the formula used for the calculation or disposal of the property in a partnership situation was changed. However, the main changes in 2006 came by use of regulations contained in the SDLT (Variation of FA 2003) Regulations 2006 (SI 2006/3237). These Regulations lead on to the introduction of s 75A of FA 2003 and also to the production of changes to partnership transactions and the introduction of a general anti-avoidance regulation (GAAR).These specific changes to s 75A are dealt with under section F below.

19.26 The main thrust of the arrangements introduced from 6 December 2006, which followed on from the pre-budget report of that year, were to block with immediate effect, a number of SDLT schemes, principally involving the use of partnerships. They also took the opportunity to introduce a wide-ranging anti-avoidance measure which in practical terms may affect innocent, commercial, property structures.

19.27 The first change was the restriction of group relief on transfers into wholly owned partnerships, unless the full conditions of company's group relief were satisfied. Therefore, at the date of transfer there must be no arrangements to sell any group partner, and the transfer money must be bona fide and not for the avoidance of SDLT, income tax, corporation tax or capital gains tax. This also introduced the possibility that any sale of a group partner within three years of the initial transaction would trigger a claw back of the relief granted.

19.28 Where property is transferred out of a group partnership to group members, no group relief would be available. The introduction of a GAAR was intended to stop schemes which sought to combine subsale relief with distributions in specie and lease variation structures. To be "caught" within the GAR provisions the following circumstances must apply.

- There is a disposal of land by vendor and a purchaser acquires it (or an interest in it).
- There are a number of connected or "scheme" transactions.
- The total SDLT payable is less than would have been payable had the vendor sold directly to the purchaser.

If these conditions are met, SDLT is imposed as if there is a straight sale of the property for the highest consideration paid under any of the scheme transactions.

19.29 Unlike other anti-avoidance legislation there is no requirement that the scheme is specifically for the avoidance or mitigation of tax, commonly called a "purpose" tax test. The only requirement is that the other transactions involved must be connected. When taken together with the very wide SDLT disclosure regulation set out above, the GAAR is likely to catch what might be described as "normal" commercial transactions.

19.30 Some commentators believe that the legislation seeks to ensure that taxpayers caught by the particular circumstances, set out above, pay the maximum conceivable tax that could be due, rather than the tax properly payable.

F. Anti-avoidance 2007

19.31 The FA 2007, at Part 5, introduced the SDLT anti-avoidance provisions previously contained in the regulations issued, with effect, from December 2006. These are set out in ss 71 and 72 of the FA 2007.

19.32 The legislation extended the scope of the transactions involved in what became s 75A of FA 2003 to also include the following at s 75A(2):

a. a non-land transaction

b. an agreement, offer or undertaking not to take specified action

c. any kind of arrangement whether or not it could not otherwise be described as a transaction

d. a transaction which takes place after the acquisition by a purchaser of the chargeable interest.

19.33 Section 71 of FA 2007 also introduced two further ss 75B and 75C to the FA 2003. These contain, in s 75B under Anti-avoidance: Incidental Transactions, and in s 75C under Anti-avoidance: Supplemental. These two sections seem to bring into consideration those transactions which, otherwise, would be incidental. However, they are not so if they form part of a process or series of transactions by which the transfer is affected.

19.34 It also brings into account the transfer of shares or securities which shall be ignored for the purposes of s 75A, except in circumstances where, but for the provisions of s 75C of FA 2003, there would be the first of a series in a scheme of transactions.

19.35 It is clear that the disclosure rules have brought to light a number of schemes of great complexity and that the provisions envisaged in the GAAR are intended to specifically sweep up schemes which have been identified by HMRC as being used for the avoidance of SDLT.

19.36 Section 72 of FA 2007 makes variations to Schedule 15 to FA 2003 in relation to partnerships. This seeks to incorporate the variations brought in by the SDLT (Variation of Finance Act 2003) Regulations 2006 with effect from December 2006. These items are dealt with in Chapter 9 under the heading of Partnerships.

G. The current situation in 2008

19.37 As set out above, the responsibility for disclosure now rests with the promoter. This provision is widely drawn and now involves any "tax advisor" unless protected by legal privilege. It also may include bodies such as banks and finance houses. Three categories of person are excluded, they are tax advisors who perform an advisory "non-designer" role, conveyancers, unless directly involved as promoters, and "spectator" tax advisors who are involved in part but not the whole scheme.

19.38 There are six steps which are excluded from forming part of a scheme, however their inclusion as a step in a scheme will not give that scheme "exemption", they are:

- the acquisition of a chargeable interest by a special purpose vehicle (SPV)
- a claim to a relief (see Chapter 1)
- sale of shares in an SPV
- not exercising an election to waive exemption from VAT (see also Chapter 8)
- transferring a business as a transfer of a going concern (TOGC) see also Chapter 8)
- undertaking a joint venture.

19.39 The scope of the GAAR will extend over time as "schemes" develop and progress. At present there is consultation into extending the GAAR into residential properties worth over £1 million and "attacking" the use of SPVs for residential avoidance schemes, particularly by overseas based persons wishing to live in the United Kingdom.

Appendix 1

Land Transaction Return

HM Revenue & Customs

Land Transaction Return

Your transaction return

How to fill in this return

The guidance notes that come with this return will help you answer the questions.

- Write inside the boxes. Use black ink and CAPITAL letters.
- If you make a mistake, please cross it out and write the correct information underneath.
- **Leave blank any boxes that don't apply to you** – please don't strike through anything irrelevant.
- Show amounts in whole pounds only, rounded down to the nearest pound. Ignore the pence.
- Fill out the payslip on page 7.
- Do not fold the return. Send it back to us unfolded in the envelope provided.
- **Photocopies are not acceptable.**

If you need help with any part of this return or with anything in the guidance notes, please phone the Stamp Taxes enquiry line on **0845 603 0135**, open 8:30am to 5:00pm Monday to Friday, except Bank Holidays. You can get further copies of this return and any supplementary returns from the Orderline on **0845 302 1472**.

Starting your return

ABOUT THE TRANSACTION

1 **Type of property**

Enter code from the guidance notes

2 **Description of transaction**

Enter code from the guidance notes

3 **Interest transferred or created**

Enter code from the guidance notes

4 **Effective date of transaction**

D D M M Y Y Y Y

5 **Any restrictions, covenants or conditions affecting the value of the interest transferred or granted?** Put 'X' in one box

Yes No

If 'yes' please provide details

6 **Date of contract or conclusion of missives**

D D M M Y Y Y Y

7 **Is any land exchanged or part-exchanged?** Put 'X' in one box

Yes No

If 'yes' please complete address of location

Postcode

House or building number

Rest of address, including house name, building name or flat number

8 **Is the transaction pursuant to a previous option agreement?** Put 'X' in one box

Yes No

SDLT 1 PG 1 IMS6/05

ABOUT THE TAX CALCULATION

9 Are you claiming relief? Put 'X' in one box

Yes No

If 'yes' please show the reason

Enter code from the guidance notes

Enter the charity's registered number, if available, or the company's CIS number

For relief claimed on part of the property only, please enter the amount remaining chargeable

£ . 0 0

10 What is the total consideration in money or money's worth, including any VAT actually payable for the transaction notified?

£ . 0 0

11 If the total consideration for the transaction includes VAT, please state the amount

£ . 0 0

12 What form does the consideration take?
Enter the relevant codes from the guidance notes

13 Is this transaction linked to any other(s)?
Put 'X' in one box

Yes No

Total consideration or value in money or money's worth, including VAT paid for all of the linked transactions

£ . 0 0

14 Total amount of tax due for this transaction

£ . 0 0

15 Total amount paid or enclosed with this notification

£ . 0 0

Does the amount paid include payment of any penalties and any interest due? Put 'X' in one box

Yes No

ABOUT LEASES

If this doesn't apply, go straight to box 26 on page 3

16 Type of lease

Enter code from the guidance notes

17 Start date as specified in lease

D D M M Y Y Y Y

18 End date as specified in lease

D D M M Y Y Y Y

19 Rent-free period
Number of months

20 Annual starting rent inclusive of VAT (actually) payable

£ . 0 0

End date for starting rent

D D M M Y Y Y Y

Later rent known? Put 'X' in one box

Yes No

21 What is the amount of VAT, if any?

£ . 0 0

22 Total premium payable

£ . 0 0

23 Net present value upon which tax is calculated

£ . 0 0

24 Total amount of tax due – premium

£ . 0 0

25 Total amount of tax due – NPV

£ . 0 0

Check the guidance notes to see if you will need to complete supplementary return 'Additional details about the transaction, including leases', SDLT4.

SDLT 1 PG 2

ABOUT THE LAND including buildings

Where more than one piece of land is being sold or you cannot complete the address field in the space provided, please complete the supplementary return 'Additional details about the land', SDLT3.

26 **Number of properties included**

27 **Where more than one property is involved, do you want a certificate for each property?** Put 'X' in one box

Yes No

28 **Address or situation of land**

Postcode

House or building number

Rest of address, including house name, building name or flat number

Is the rest of the address on the supplementary return 'Additional details about the land', SDLT3?
Put 'X' in one box

Yes No

29 **Local authority number**

30 **Title number, if any**

31 **NLPG UPRN**

32 **If agricultural or development land, what is the area (if known)?** Put 'X' in one box

Hectares Square metres

Area

33 **Is a plan attached?** Please note that the form reference number should be written/displayed on map. Put 'X' in one box

Yes No

ABOUT THE VENDOR including transferor, lessor

34 **Number of vendors included** (Note: if more than one vendor, complete boxes 45 to 48)

35 **Title** Enter MR, MRS, MISS, MS or other title
Note: only complete for an individual

36 **Vendor (1) surname or company name**

37 **Vendor (1) first name(s)** Note: only complete for an individual

38 **Vendor (1) address**

Postcode

House or building number

Rest of address, including house name, building name or flat number

SDLT 1 PG 3

ABOUT THE VENDOR CONTINUED

39 **Agent's name**

40 **Agent's address**
Postcode

Building number

Rest of address, including building name

41 **Agent's DX number and exchange**

42 **Agent's e-mail address**

43 **Agent's reference**

44 **Agent's telephone number**

ADDITIONAL VENDOR

Details of other people involved (including transferor, lessor), other than vendor (1). If more than one additional vendor please complete supplementary return 'Land Transaction Return - Additional vendor/purchaser details', SDLT2.

45 **Title** Enter MR, MRS, MISS, MS or other title
Note: only complete for an individual

46 **Vendor (2) surname or company name**

47 **Vendor (2) first name(s)**
Note: only complete for an individual

48 **Vendor (2) address**

Put 'X' in this box if the same as box 38.

If not, please give address below
Postcode

House or building number

Rest of address, including house name, building name or flat number

ABOUT THE PURCHASER including transferee, lessee

49 **Number of purchasers included** (Note: if more than one purchaser is involved, complete boxes 65 to 69)

50 **National Insurance number (purchaser 1), if you have one.** Note: only complete for an individual

51 **Title** Enter MR, MRS, MISS, MS or other title
Note: only complete for an individual

52 **Purchaser (1) surname or company name**

53 **Purchaser (1) first name(s)**
Note: only complete for an individual

54 **Purchaser (1) address**

Put 'X' in this box if the same address as box 28.

If not, please give address below

Postcode

House or building number

Rest of address, including house name, building name or flat number

55 **Is the purchaser acting as a trustee?** Put 'X' in one box

Yes No

56 **Please give a daytime telephone number – this will help us if we need to contact you about your return**

57 **Are the purchaser and vendor connected?**
Put 'X' in one box

Yes No

58 **To which address shall we send the certificate?**
Put 'X' in one box

Property (box 28) Purchaser's (box 54)

Agent's (box 61)

59 **I authorise my agent to handle correspondence on my behalf.** Put 'X' in one box

Yes No

60 **Agent's name**

61 **Agent's address**
Postcode

Building number

Rest of address, including building name

62 **Agent's DX number and exchange**

63 **Agent's reference**

64 **Agent's telephone number**

SDLT 1 PG 5

ADDITIONAL PURCHASER

Details of other people involved (including transferee, lessee), other than purchaser (1). If more than one additional purchaser, please complete supplementary return 'Land Transaction Return - Additional vendor/purchaser details', SDLT2.

65 **Title** Enter MR, MRS, MISS, MS or other title
Note: only complete for an individual

66 **Purchaser (2) surname or company name**

67 **Purchaser (2) first name(s)**
Note: only complete for an individual

68 **Purchaser (2) address**

Put 'X' in this box if the same as purchaser (1) (box 54).

If not, please give address below

Postcode

House or building number

Rest of address, including house name, building name or flat number

69 **Is purchaser (2) acting as a trustee?** Put 'X' in one box

Yes No

ADDITIONAL SUPPLEMENTARY RETURNS

70 **How many supplementary returns have you enclosed with this return?** Write the number in each box. If none, please put '0'.

Additional vendor/purchaser details, SDLT2

Additional details about the land, SDLT3

Additional details about the transaction, including leases, SDLT4

DECLARATION

71 **The purchaser(s) must sign this return.** Read the guidance notes in booklet SDLT6, in particular the section headed *Who should complete and sign the Land Transaction Return?'*.

If you give false information, you may face financial penalties and prosecution.
The information I have given on this return is correct and complete to the best of my knowledge and belief.

Signature of purchaser 1 Signature of purchaser 2

Please keep a copy of this return and a note of the unique transaction reference number, which is in the 'Reference' box on the payslip.

Finally, please send your completed return to:
HM Revenue & Customs, Stamp Taxes/SDLT, Comben House, Farriers Way, NETHERTON, Merseyside, Great Britain, L30 4RN, or the DX address is: Rapid Data Capture Centre, DX725593, Bootle 9

Please don't fold it - keep it flat and use the envelope provided. Fill out the payslip on the next page and pay in accordance with the 'How to pay' instructions.

SDLT 1 PG 6

How to pay

Please allow enough time for payment to reach us by the due date. We suggest you allow at least 3 working days for this.

MOST SECURE AND EFFICIENT

We recommend the following payment methods. These are the most secure and efficient.

1. Direct Payment
Use the Internet or telephone to make payment. Provide your bank or building society with the following information
- payment account
- sort code 10-50-41
- account number 23456000
- your reference as shown on the payslip.

2. BillPay
You can pay by Debit Card over the Internet. Visit **www.billpayment.co.uk/hmrc** and follow the guidance.

3. At your bank
Take this form with payment to **your** bank and where possible to **your own** branch. Other banks may refuse to accept payment. If paying by cheque, make your cheque payable to 'HM REVENUE & CUSTOMS ONLY'.

4. At a Post Office
Take this form with your payment to any Post Office. If paying by cheque, make your cheque payable to 'POST OFFICE LTD'. The Post Office also accept payment by Debit Card.

5. Alliance & Leicester Commercial Bank Account
Alliance & Leicester Commercial Bank customers can instruct their bank to arrange payment.

OTHER PAYMENT METHODS

By post
If you use this method
- Make your cheque payable to 'HM REVENUE & CUSTOMS ONLY'.
- Write your payslip reference after 'HM REVENUE & CUSTOMS ONLY'.
- Send the payslip and your cheque, **both unfolded**, in the envelope provided to
  HM Revenue & Customs SDLT
  Netherton
  Merseyside
  L30 4RN

By DX
As above, but send to
Rapid Data Capture Centre
DX725593
Bootle 9

FURTHER PAYMENT INFORMATION

You can find further payment information at **www.hmrc.gov.uk/howtopay**

Alliance & Leicester Transcash COMMERCIAL BANK
Bootle Merseyside GIR 0AA

Payslip HM Revenue & Customs

bank giro credit

159
209
24

Reference

Credit account number
610 5041

£
Amount due
(not payable at PO counter)
CHEQUE ACCEPTABLE

By transfer from Alliance & Leicester account number

For official use only

Name

Cashier's stamp and initials

Signature ____________ Date ____________

CASH
CHEQUE
£

BANK OF ENGLAND
HEAD OFFICE COLLECTION A/C
HM REVENUE & CUSTOMS

10-50-41

SDLT1/P

Please do not fold this payslip or write or mark below this line

Appendix 2

YP at 3.5% and PV at 3.5%

No Income Tax **Single Rate**

YEARS' PURCHASE

	Rate Per Cent								
Yrs.	**2**	**2.25**	**2.5**	**2.75**	**3**	**3.25**	**3.5**	**3.75**	Yrs.
1	0.9804	0.9780	0.9756	0.9732	0.9709	0.9685	0.9662	0.9639	**1**
2	1.9416	1.9345	1.9274	1.9204	1.9135	1.9066	1.8997	1.8929	**2**
3	2.8839	2.8699	2.8560	2.8423	2.8286	2.8151	2.8016	2.7883	**3**
4	3.8077	3.7847	3.7620	3.7394	3.7171	3.6950	3.6731	3.6514	**4**
5	4.7135	4.6795	4.6458	4.6126	4.5797	4.5472	4.5151	4.4833	**5**
6	5.6014	5.5545	5.5081	5.4624	5.4172	5.3726	5.3286	5.2851	**6**
7	6.4720	6.4102	6.3494	6.2894	6.2303	6.1720	6.1145	6.0579	**7**
8	7.3255	7.2472	7.1701	7.0943	7.0197	6.9462	6.8740	6.8028	**8**
9	8.1622	8.0657	7.9709	7.8777	7.7861	7.6961	7.6077	7.5208	**9**
10	8.9826	8.8662	8.7521	8.6401	8.5302	8.4224	8.3166	8.2128	**10**
11	9.7868	9.6491	9.5142	9.3821	9.2526	9.1258	9.0016	8.8798	**11**
12	10.5753	10.4148	10.2578	10.1042	9.9540	9.8071	9.6633	9.5227	**12**
13	11.3484	11.1636	10.9832	10.8070	10.6350	10.4669	10.3027	10.1424	**13**
14	12.1062	11.8959	11.6909	11.4910	11.2961	11.1060	10.9205	10.7396	**14**
15	12.8493	12.6122	12.3814	12.1567	11.9379	11.7249	11.5174	11.3153	**15**
16	13.5777	13.3126	13.0550	12.8046	12.5611	12.3244	12.0941	11.8702	**16**
17	14.2919	13.9977	13.7122	13.4351	13.1661	12.9049	12.6513	12.4050	**17**
18	14.9920	14.6677	14.3534	14.0488	13.7535	13.4673	13.1897	12.9205	**18**
19	15.6785	15.3229	14.9789	14.6460	14.3238	14.0119	13.7098	13.4173	**19**
20	16.3514	15.9637	15.5892	15.2273	14.8775	14.5393	14.2124	13.8962	**20**
21	17.0112	16.5904	16.1845	15.7929	15.4150	15.0502	14.6980	14.3578	**21**
22	17.6580	17.2034	16.7654	16.3435	15.9369	15.5450	15.1671	14.8027	**22**
23	18.2922	17.8028	17.3321	16.8793	16.4436	16.0242	15.6204	15.2315	**23**
24	18.9139	18.3890	17.8850	17.4008	16.9355	16.4883	16.0584	15.6448	**24**
25	19.5235	18.9624	18.4244	17.9083	17.4131	16.9379	16.4815	16.0432	**25**
26	20.1210	19.5231	18.9506	18.4023	17.8768	17.3732	16.8904	16.4272	**26**
27	20.7069	20.0715	19.4640	18.8830	18.3270	17.7949	17.2854	16.7973	**27**
28	21.2813	20.6078	19.9649	19.3508	18.7641	18.2033	17.6670	17.1540	**28**
29	21.8444	21.1323	20.4535	19.8062	19.1885	18.5988	18.0358	17.4978	**29**
30	22.3965	21.6453	20.9303	20.2493	19.6004	18.9819	18.3920	17.8292	**30**
31	22.9377	22.1470	21.3954	20.6806	20.0004	19.3529	18.7363	18.1487	**31**
32	23.4683	22.6377	21.8492	21.1003	20.3888	19.7123	19.0689	18.4565	**32**
33	23.9886	23.1175	22.2919	21.5088	20.7658	20.0603	19.3902	18.7533	**33**
34	24.4986	23.5868	22.7238	21.9064	21.1318	20.3974	19.7007	19.0393	**34**
35	24.9986	24.0458	23.1452	22.2933	21.4872	20.7239	20.0007	19.3150	**35**
36	25.4888	24.4947	23.5563	22.6699	21.8323	21.0401	20.2905	19.5807	**36**
37	25.9695	24.9337	23.9573	23.0364	22.1672	21.3463	20.5705	19.8369	**37**
38	26.4406	25.3630	24.3486	23.3931	22.4925	21.6429	20.8411	20.0837	**38**
39	26.9026	25.7829	24.7303	23.7402	22.8082	21.9302	21.1025	20.3217	**39**
40	27.3555	26.1935	25.1028	24.0781	23.1148	22.2084	21.3551	20.5510	**40**
41	27.7995	26.5951	25.4661	24.4069	23.4124	22.4779	21.5991	20.7720	**41**
42	28.2348	26.9879	25.8206	24.7269	23.7014	22.7389	21.8349	20.9851	**42**
43	28.6616	27.3720	26.1664	25.0384	23.9819	22.9917	22.0627	21.1905	**43**
44	29.0800	27.7477	26.5038	25.3415	24.2543	23.2365	22.2828	21.3884	**44**
45	29.4902	28.1151	26.8330	25.6365	24.5187	23.4736	22.4955	21.5792	**45**
46	29.8923	28.4744	27.1542	25.9236	24.7754	23.7032	22.7009	21.7631	**46**
47	30.2866	28.8259	27.4675	26.2030	25.0247	23.9256	22.8994	21.9403	**47**
48	30.6731	29.1695	27.7732	26.4749	25.2667	24.1411	23.0912	22.1111	**48**
49	31.0521	29.5057	28.0714	26.7396	25.5017	24.3497	23.2766	22.2758	**49**
50	31.4236	29.8344	28.3623	26.9972	25.7298	24.5518	23.4556	22.4345	**50**

40

No Income Tax **Single Rate**

YEARS' PURCHASE

Rate Per Cent

Yrs.	2	2.25	2.5	2.75	3	3.25	3.5	3.75	Yrs.
51	31.7878	30.1559	28.6462	27.2479	25.9512	24.7475	23.6286	22.5875	**51**
52	32.1449	30.4703	28.9231	27.4918	26.1662	24.9370	23.7958	22.7349	**52**
53	32.4950	30.7778	29.1932	27.7293	26.3750	25.1206	23.9573	22.8770	**53**
54	32.8383	31.0785	29.4568	27.9604	26.5777	25.2984	24.1133	23.0140	**54**
55	33.1748	31.3727	29.7140	28.1853	26.7744	25.4706	24.2641	23.1460	**55**
56	33.5047	31.6603	29.9649	28.4042	26.9655	25.6374	24.4097	23.2733	**56**
57	33.8281	31.9416	30.2096	28.6172	27.1509	25.7989	24.5504	23.3959	**57**
58	34.1452	32.2167	30.4484	28.8245	27.3310	25.9554	24.6864	23.5141	**58**
59	34.4561	32.4858	30.6814	29.0263	27.5058	26.1069	24.8178	23.6281	**59**
60	34.7609	32.7490	30.9087	29.2227	27.6756	26.2537	24.9447	23.7379	**60**
61	35.0597	33.0063	31.1304	29.4138	27.8404	26.3958	25.0674	23.8438	**61**
62	35.3526	33.2580	31.3467	29.5998	28.0003	26.5335	25.1859	23.9458	**62**
63	35.6398	33.5042	31.5578	29.7808	28.1557	26.6668	25.3004	24.0442	**63**
64	35.9214	33.7449	31.7637	29.9570	28.3065	26.7959	25.4110	24.1389	**64**
65	36.1975	33.9803	31.9646	30.1285	28.4529	26.9210	25.5178	24.2303	**65**
66	36.4681	34.2106	32.1606	30.2953	28.5950	27.0421	25.6211	24.3184	**66**
67	36.7334	34.4358	32.3518	30.4578	28.7330	27.1594	25.7209	24.4032	**67**
68	36.9936	34.6560	32.5383	30.6158	28.8670	27.2731	25.8173	24.4851	**68**
69	37.2486	34.8714	32.7203	30.7697	28.9971	27.3831	25.9104	24.5639	**69**
70	37.4986	35.0821	32.8979	30.9194	29.1234	27.4897	26.0004	24.6399	**70**
71	37.7437	35.2881	33.0711	31.0651	29.2460	27.5929	26.0873	24.7132	**71**
72	37.9841	35.4896	33.2401	31.2069	29.3651	27.6929	26.1713	24.7838	**72**
73	38.2197	35.6866	33.4050	31.3449	29.4807	27.7897	26.2525	24.8518	**73**
74	38.4507	35.8794	33.5658	31.4792	29.5929	27.8835	26.3309	24.9174	**74**
75	38.6771	36.0678	33.7227	31.6100	29.7018	27.9744	26.4067	24.9807	**75**
76	38.8991	36.2522	33.8758	31.7372	29.8076	28.0623	26.4799	25.0416	**76**
77	39.1168	36.4324	34.0252	31.8610	29.9103	28.1475	26.5506	25.1003	**77**
78	39.3302	36.6087	34.1709	31.9815	30.0100	28.2301	26.6190	25.1569	**78**
79	39.5394	36.7812	34.3131	32.0988	30.1068	28.3100	26.6850	25.2115	**79**
80	39.7445	36.9498	34.4518	32.2129	30.2008	28.3874	26.7488	25.2641	**80**
81	39.9456	37.1147	34.5871	32.3240	30.2920	28.4624	26.8104	25.3148	**81**
82	40.1427	37.2760	34.7192	32.4321	30.3806	28.5350	26.8700	25.3637	**82**
83	40.3360	37.4337	34.8480	32.5374	30.4666	28.6053	26.9275	25.4108	**83**
84	40.5255	37.5880	34.9736	32.6398	30.5501	28.6734	26.9831	25.4562	**84**
85	40.7113	37.7389	35.0962	32.7394	30.6312	28.7394	27.0368	25.4999	**85**
86	40.8934	37.8864	35.2158	32.8364	30.7099	28.8033	27.0887	25.5421	**86**
87	41.0720	38.0307	35.3325	32.9308	30.7863	28.8652	27.1388	25.5827	**87**
88	41.2470	38.1719	35.4463	33.0227	30.8605	28.9251	27.1873	25.6219	**88**
89	41.4187	38.3099	35.5574	33.1121	30.9325	28.9831	27.2341	25.6597	**89**
90	41.5869	38.4449	35.6658	33.1992	31.0024	29.0394	27.2793	25.6961	**90**
91	41.7519	38.5769	35.7715	33.2838	31.0703	29.0938	27.3230	25.7312	**91**
92	41.9136	38.7060	35.8746	33.3663	31.1362	29.1466	27.3652	25.7650	**92**
93	42.0722	38.8323	35.9752	33.4465	31.2002	29.1976	27.4060	25.7976	**93**
94	42.2276	38.9558	36.0734	33.5246	31.2623	29.2471	27.4454	25.8290	**94**
95	42.3800	39.0766	36.1692	33.6006	31.3227	29.2950	27.4835	25.8592	**95**
96	42.5294	39.1947	36.2626	33.6745	31.3812	29.3414	27.5203	25.8884	**96**
97	42.6759	39.3102	36.3538	33.7465	31.4381	29.3864	27.5558	25.9166	**97**
98	42.8195	39.4232	36.4427	33.8165	31.4933	29.4299	27.5902	25.9437	**98**
99	42.9603	39.5337	36.5295	33.8847	31.5469	29.4720	27.6234	25.9698	**99**
100	43.0984	39.6417	36.6141	33.9510	31.5989	29.5129	27.6554	25.9950	**100**
PERP	50.0000	44.4444	40.0000	36.3636	33.3333	30.7692	28.5714	26.6667	**PERP**

No Income Tax

PRESENT VALUE OF £1

	Rate Per Cent					
Yrs.	**3.25**	**3.5**	**3.75**	**4**	**4.25**	Yrs.
1	.9685230	.9661836	.9638554	.9615385	.9592326	**1**
2	.9380368	.9335107	.9290173	.9245562	.9201272	**2**
3	.9085102	.9019427	.8954383	.8889964	.8826160	**3**
4	.8799130	.8714422	.8630731	.8548042	.8466341	**4**
5	.8522160	.8419732	.8318777	.8219271	.8121190	**5**
6	.8253908	.8135006	.8018098	.7903145	.7790111	**6**
7	.7994100	.7859910	.7728287	.7599178	.7472528	**7**
8	.7742470	.7594116	.7448952	.7306902	.7167893	**8**
9	.7498760	.7337310	.7179712	.7025867	.6875676	**9**
10	.7262722	.7089188	.6920205	.6755642	.6595373	**10**
11	.7034113	.6849457	.6670077	.6495809	.6326497	**11**
12	.6812700	.6617833	.6428990	.6245970	.6068582	**12**
13	.6598257	.6394042	.6196617	.6005741	.5821182	**13**
14	.6390564	.6177818	.5972643	.5774751	.5583868	**14**
15	.6189408	.5968906	.5756764	.5552645	.5356228	**15**
16	.5994584	.5767059	.5548688	.5339082	.5137868	**16**
17	.5805892	.5572038	.5348133	.5133732	.4928411	**17**
18	.5623140	.5383611	.5154827	.4936281	.4727493	**18**
19	.5446141	.5201557	.4968508	.4746424	.4534765	**19**
20	.5274713	.5025659	.4788923	.4563869	.4349895	**20**
21	.5108680	.4855709	.4615830	.4388336	.4172561	**21**
22	.4947874	.4691506	.4448993	.4219554	.4002456	**22**
23	.4792130	.4532856	.4288186	.4057263	.3839287	**23**
24	.4641288	.4379571	.4133191	.3901215	.3682769	**24**
25	.4495195	.4231470	.3983799	.3751168	.3532632	**25**
26	.4353699	.4088377	.3839806	.3606892	.3388616	**26**
27	.4216658	.3950122	.3701018	.3468166	.3250471	**27**
28	.4083930	.3816543	.3567246	.3334775	.3117958	**28**
29	.3955380	.3687482	.3438309	.3206514	.2990847	**29**
30	.3830877	.3562784	.3314033	.3083187	.2868918	**30**
31	.3710292	.3442303	.3194249	.2964603	.2751959	**31**
32	.3593503	.3325897	.3078794	.2850579	.2639769	**32**
33	.3480391	.3213427	.2967512	.2740942	.2532153	**33**
34	.3370839	.3104761	.2860253	.2635521	.2428923	**34**
35	.3264735	.2999769	.2756870	.2534155	.2329903	**35**
36	.3161971	.2898327	.2657224	.2436687	.2234919	**36**
37	.3062441	.2800316	.2561180	.2342968	.2143807	**37**
38	.2966045	.2705619	.2468607	.2252854	.2056409	**38**
39	.2872683	.2614125	.2379380	.2166206	.1972575	**39**
40	.2782259	.2525725	.2293379	.2082890	.1892158	**40**
41	.2694682	.2440314	.2210486	.2002779	.1815020	**41**
42	.2609862	.2357791	.2130588	.1925749	.1741026	**42**
43	.2527711	.2278059	.2053579	.1851682	.1670049	**43**
44	.2448146	.2201023	.1979353	.1780463	.1601966	**44**
45	.2371086	.2126592	.1907811	.1711984	.1536658	**45**
46	.2296451	.2054679	.1838854	.1646139	.1474012	**46**
47	.2224166	.1985197	.1772389	.1582826	.1413921	**47**
48	.2154156	.1918065	.1708327	.1521948	.1356279	**48**
49	.2086349	.1853202	.1646580	.1463411	.1300987	**49**
50	.2020677	.1790534	.1587065	.1407126	.1247949	**50**

No Income Tax

PRESENT VALUE OF £1

	Rate Per Cent					
Yrs.	3.25	3.5	3.75	4	4.25	Yrs.
1	.9685230	.9661836	.9638554	.9615385	.9592326	1
2	.9380368	.9335107	.9290173	.9245562	.9201272	2
3	.9085102	.9019427	.8954383	.8889964	.8826160	3
4	.8799130	.8714422	.8630731	.8548042	.8466341	4
5	.8522160	.8419732	.8318777	.8219271	.8121190	5
6	.8253908	.8135006	.8018098	.7903145	.7790111	6
7	.7994100	.7859910	.7728287	.7599178	.7472528	7
8	.7742470	.7594116	.7448952	.7306902	.7167893	8
9	.7498760	.7337310	.7179712	.7025867	.6875676	9
10	.7262722	.7089188	.6920205	.6755642	.6595373	10
11	.7034113	.6849457	.6670077	.6495809	.6326497	11
12	.6812700	.6617833	.6428990	.6245970	.6068582	12
13	.6598257	.6394042	.6196617	.6005741	.5821182	13
14	.6390564	.6177818	.5972643	.5774751	.5583868	14
15	.6189408	.5968906	.5756764	.5552645	.5356228	15
16	.5994584	.5767059	.5548688	.5339082	.5137868	16
17	.5805892	.5572038	.5348133	.5133732	.4928411	17
18	.5623140	.5383611	.5154827	.4936281	.4727493	18
19	.5446141	.5201557	.4968508	.4746424	.4534765	19
20	.5274713	.5025659	.4788923	.4563869	.4349895	20
21	.5108680	.4855709	.4615830	.4388336	.4172561	21
22	.4947874	.4691506	.4448993	.4219554	.4002456	22
23	.4792130	.4532856	.4288186	.4057263	.3839287	23
24	.4641288	.4379571	.4133191	.3901215	.3682769	24
25	.4495195	.4231470	.3983799	.3751168	.3532632	25
26	.4353699	.4088377	.3839806	.3606892	.3388616	26
27	.4216658	.3950122	.3701018	.3468166	.3250471	27
28	.4083930	.3816543	.3567246	.3334775	.3117958	28
29	.3955380	.3687482	.3438309	.3206514	.2990847	29
30	.3830877	.3562784	.3314033	.3083187	.2868918	30
31	.3710292	.3442303	.3194249	.2964603	.2751959	31
32	.3593503	.3325897	.3078794	.2850579	.2639769	32
33	.3480391	.3213427	.2967512	.2740942	.2532153	33
34	.3370839	.3104761	.2860253	.2635521	.2428923	34
35	.3264735	.2999769	.2756870	.2534155	.2329903	35
36	.3161971	.2898327	.2657224	.2436687	.2234919	36
37	.3062441	.2800316	.2561180	.2342968	.2143807	37
38	.2966045	.2705619	.2468607	.2252854	.2056409	38
39	.2872683	.2614125	.2379380	.2166206	.1972575	39
40	.2782259	.2525725	.2293379	.2082890	.1892158	40
41	.2694682	.2440314	.2210486	.2002779	.1815020	41
42	.2609862	.2357791	.2130588	.1925749	.1741026	42
43	.2527711	.2278059	.2053579	.1851682	.1670049	43
44	.2448146	.2201023	.1979353	.1780463	.1601966	44
45	.2371086	.2126592	.1907811	.1711984	.1536658	45
46	.2296451	.2054679	.1838854	.1646139	.1474012	46
47	.2224166	.1985197	.1772389	.1582826	.1413921	47
48	.2154156	.1918065	.1708327	.1521948	.1356279	48
49	.2086349	.1853202	.1646580	.1463411	.1300987	49
50	.2020677	.1790534	.1587065	.1407126	.1247949	50

92

Appendix 3

Connected Persons

Connected persons

Section 839 of Income and Corporation Taxes Act 1988

1. For the purposes of, and subject to the provisions of the Tax Acts which apply this section, any question whether a person is connected with another shall be determined in accordance with the following provisions of this section (any provision that one person is connected with another being taken to mean that they are connected with one another).

2. A person is connected with an individual if that person is the individual's wife or husband, or is a relative, or the wife or husband of a relative of the individual or of the individual's wife or husband.

3. A person, in their capacity as trustee of a settlement, is connected with:

   a. any individual who in relation to the settlement is a settlor
   b. any person who is connected with such an individual
   c. any body corporate, which is connected with that settlement.

   In this subsection "settlement" and "settlor" have the same meaning as in Chapter IA of Part XV (see s 660G(1) and (2)).

3A. For the purpose of the subsection (3) above, a body corporate is connected with a settlement if:

a. it is a close company (or only not a close company because it is not resident in the United Kingdom) and the participators include the trustees of the settlement

b. it is controlled (within the meaning of s 840) by a company falling within paragraph (a) above.

4. Except in relation to acquisitions or disposals of partnership assets pursuant to bona fide commercial arrangement, a person is connected with any person with whom they are in partnership, and with the wife or husband or relative of any individual with whom they are in partnership.

5. A company is connected with another company:

a. if the same person has control of both, or a person has control of one and the person connected with them, or they and the person connected with them have control of the other

b. if a group of two or more persons has control of each company, and the groups either consist of the same persons or could be regarded as consisting of the same persons by treating (in one or more cases) a member of either group as replaced by a person with whom they are connected.

6. A company is connected with another person if that person has control of it or if that person and persons connected with them together have control of it.

7. Any two or more persons acting together to secure or exercise control of a company shall be treated in relation to that company as connected with one another and with any person acting on the directions of any of them to secure or exercise control of the company.

8. In this section:

- "company" includes any body corporate or unincorporated association, but does not include a partnership, and this section shall apply in relation to any unit trust scheme as if the scheme were a company and as if the rights of the unit holders were shares in the company

- "control" shall be construed in accordance with s 416
- "relative" means brother, sister, ancestor or lineal descendant.

Section 840

For the purposes of, and subject to, the provisions of the Tax Acts which apply this section, "control", in relation to a body corporate, means the power of a person to secure:

a. by means of the holding of shares or the possession of voting power in or in relation to that or any other body corporate; or
b. by virtue of any powers conferred by the articles of association or other document regulating that or any other body corporate;

that the affairs of the first-mentioned body corporate are conducted in accordance with the wishes of that person, and, in relation to a partnership, means the right to a share of more than one-half of the assets, or of more than one-half of the income of the partnership.

Many of the above are re-stated in the Income Tax Act 2007.

Appendix 4

Group Relief Reconstruction Relief Acquisition Relief

Schedule 7 of Finance Act 2003 — SDLT: Group relief, reconstruction and acquisition reliefs

1. Schedule 7 to the Finance Act 2003 (FA 2003) is set out in two parts. Part 1 deals with group relief, restrictions on group relief, withdrawal of group relief, where group relief is not withdrawn, recovery of tax following withdrawal of relief from another group company or director and supplementary provisions. Part 2, which deals with reconstruction and acquisition reliefs, follows a similar pattern. In common with much of stamp duty land tax (SDLT) legislation, Schedule 7 deals with the basic reliefs relatively briefly and with potential avoidance by exploitation of the reliefs in great detail.

2. The regulations are complex and, as an alternative to setting them out in full, they have been dealt with in a précis form. It is recommended, where a situation arises which may, potentially, lead to the withdrawal of reliefs, under Schedule 7 to FA 2003, that the legislation be consulted in detail. In many Finance Acts

since the 2003 amendments, changes have been made, mainly to prevent avoidance. For example, significant changes were made in the Finance Act 2005. These were primarily anti-avoidance provisions to attack what are known as "drop and bounce" or "drop and slide" transactions.

A. Group relief

3. A transaction is exempt from SDLT (Schedule 7 to FA 2003, para 1) if it is between companies that are, at the effective date, in the same group. Companies are in the same group if they pass the "75%" test.

   - Companies are in the same group if one is a 75% (or more) shareholder (or owner) of the other.

   - If one is a 75% subsidiary of the other or both are 75% subsidiaries of a third company, they are also in the same group.

4. The 75% test turns on the following. Where A owns 75% of the shares of B, and is entitled to 75% of any profits on distribution by B and would be entitled to 75% of the assets of B on its winding up, the test is passed. In company law there may be a succession of subsidiary companies. In this respect the "senior" company is the one "above" in the subsidiary chain. Therefore, where A owns 75% of B which owns 75% of C, B is subsidiary to A and C is subsidiary to both B and A which are considered "above" it.

5. A can hold the share holdings in B, either entirely, or through other companies in the same group.

6. Group relief is not available (Schedule 7 to FA 2003, Para 2) if at the effective date (for SDLT) another person or persons could obtain control of the purchaser, the one liable for payment of SDLT, but not of the vendor.

7. There are specific exceptions to this rule relating to proposed share acquisitions where s 75 of Finance Act 1986 applies (Stamp Duty Acquisition Relief), or the purchaser becomes a member of the group.

8. Group relief is not available where:

- arrangements are in place to provide the purchase consideration by a person other than a group member
- the vendor and purchaser are to cease to be members of the same group by failing the 75% test.

These regulations are to prevent a company joining and leaving a group purely as a measure to avoid SDLT on a property acquisition.

9. Where, in an exempt transaction, the purchaser ceases to be a group member (with the vendor) within three years, or because of arrangements made before the transaction, relief may be withdrawn and SDLT charged on any property transaction provided that, when the purchaser ceases to be part of the vendor's group, it, or an associated company, holds a chargeable interest acquired under the relevant transaction or derived from the chargeable interest (Schedule 7 to FA 2003, para 3).

10. An exception applies where the interest has been obtained at market value and available group relief was not claimed. In such circumstances then the provisions of Schedule 7 to FA 2003, para 3, do not apply.

11. Where relief is withdrawn, the SDLT payable is based on the market value of the relevant asset (Schedule 7 of FA 2003, para 3(2)) or, alternatively, on an appropriate proportion thereof.

12. Group relief is not withdrawn (Schedule 7 of FA 2003, para 4) under para 3 where:

    - the purchaser ceases to be in the vendor's group because the vendor leaves the group
    - the vendor leaves the group due to a transaction either in its shares, or in those of another group company
    - winding up either of the vendor, or a company "above" the vendor in the subsidiary chain, causes the vendor to leave the group
    - a share acquisition causes the vendor to leave the group but s 75 of FA 1986, Stamp Duty Acquisition Relief, applies (effectively the vendor becomes grouped with the acquiring company for this purpose).

13. However if the circumstances, as detailed in para 9 above, occur, then the provisions, as set out in para 11 above, apply (Schedule 7 to FA 2003, para 4).

14. In February 2006, Her Majesty's Revenue and Customs (HMRC) published guidance on transactions where it accepted that group relief is not to be denied after all transfers (see below), being to group companies.

- The transfer of property where there is an intention to sell shares three years after transfer date.
- The transfer of property with an intention to sell shares within three years (leading to claw back relief) in order to shelter any increase in value from SDLT.
- The transfer of a property with the possibility that either of the above might occur.
- The transfer of a property, prior to the sales of shares in the transferor to prevent the property passing to the purchaser of the shares.
- The transfer of a property to offset commercially generated rental income against losses generated form a Schedule A business.
- Similarly using a property transfer to match chargeable gains against allowable losses.
- The transfer of property to a non-resident group company to avoid future capital gains.
- Transactions forming part of a commercial securitisation.
- Transferring a reversion to merge freehold and leasehold interests to avoid the capital gains tax wasting asset provisions.
- Transferring property to offset interest payable against commercially generated rental income.

15. If SDLT is chargeable under para 3 and the amount has been determined (see Chapter 4) and the tax remains unpaid six months after the due date, the following bodies or persons may be required to pay the unpaid tax.

- The vendor.
- Any company in the same group but "above" it in the subsidiary chain, as the purchaser, at the effective date.
- Any controlling director of the purchaser, or any controlling company (as at the effective date) (Schedule 7 to FA 2003, para 5).

This ensures that winding up or dissolving the company which purchased the asset, or the leaving of the purchasing company without sufficient funds, does not lead to a loss of SDLT.

16. Schedule 7 to FA 2003, para 6, allows HMRC to serve notice on any of the above within 30 days of the notice to pay. However, this notice must be served within three years of the final determination. The notice must state the amount due, and be served on the appropriate person. Service of the notice allows HMRC to pursue the unpaid tax as a debt, but also allows the payer to recover the debt from the purchaser. The amount to be paid is not allowed as a tax deduction by any of the parties to the transaction.

17. There are two changes to the group relief rules in Part 1 of Schedule7 to FA 2003. The first is in para (7) (with consequential amendments in paras (4), (5) and (8)), the second is in para (6).

First, a new para 4A is inserted into para 7, this applies where:

- a transferee claims group relief in respect of a transaction (the "relevant transaction") and within three years
- there is a change of control of the transferee, control being defined as in s 416 of the Income and Corporation Taxes Act 1988 (ICTA 1988), this change of control is called a triggering event group relief would not otherwise be withdrawn under para 3 of Schedule 7.

18. In that case, the earliest "previous transaction" must be identified. Group relief is then withdrawn if it would have been withdrawn under para 3 of Schedule 7 to FA 2003 if the vendor under the relevant transaction had been the vendor under the earliest previous transaction.

A "previous transaction" for this purpose is one where:

- group relief, reconstruction relief or acquisition relief was claimed
- the effective date of the previous transaction was less than three years before the triggering event
- the chargeable interest acquired under the relevant transaction is the same as, comprises, forms part of, or is derived from, the interest acquired under the previous transaction.

19. However, a transaction is not a "previous transaction" if since then there has been a transaction in the interest acquired under that transaction which was not exempt from charge by virtue of group relief, reconstruction relief or acquisition relief.

20. Where two or more transaction effected at the same time form the earliest "previous transaction" (for example where two chargeable interests merge to form the chargeable interest which is the subject of the relevant transaction), the reference to "vendor" in relation to the earliest "previous transaction" is taken to be a reference to the vendors in each of the earliest "previous transactions".

Examples:

- On 1 January 2006, A transfers Blackacre to B.
- On 1 January 2007, B transfers Blackacre to C.
- On 1 January 2008, C transfers Blackacre to D.
- On 1 January 2009, D transfers Blackacre to E.
- On 1 July 2009, there is a change of control of E.

21. All the transfers benefit from group relief, it may be that this causes withdrawal of the group relief claimed by E under existing legislation. If not, the earliest "previous transaction" is the B to C transfer (since the A to B transfer takes place more than three years before the triggering event) and so the group relief claimed by E is withdrawn if E had ceased to be a member of the same group as B (subject to the cases in para 4 of Schedule 7 to FA 2003 where group relief is not withdrawn, taking B as vendor).

22. Take the same facts as above except that the transfer from C to D did not benefit from group relief. The new measure has no effect here since the non-exempt C to D transfer prevents the B to C transfer being a "previous transaction" and the C to D transfer being non-exempt cannot itself be a "previous transfer".

The above measures took effect where the effective date of the relevant transaction is on or after 17 March 2005.

23. The second change was to para 4(3) of Schedule 2 to FA 2003. Paragraph 4 lists cases where group relief is not withdrawn, even though para 3 (withdrawal of group relief) applies. Paragraph 4(3) provides that group relief is not withdrawn if the vendor "leaves the group" and defines what this means. The existing legislation says that the vendor "leaves the group" if the vendor and purchaser cease to be members of the same group as the purchaser by reason of a transaction relating either to shares in the vendor or to shares in another company that as a result of the transaction ceases to be a member of the same group as the

purchaser. The measure restricts the second option so that the other company must be above the vendor in the group structure. This change has effect where the effective date of the transaction, to which para 3 of Schedule 7 to FA 2003 applies, is on or after 17 March 2005.

B. Reconstruction and acquisition reliefs

24. If, under Schedule 7 to FA 2003, para 7, a company acquires the whole or part of another company as part of a scheme of that company's reconstruction and:

- consideration is by way of non-redeemable shares in the purchaser, either wholly or partly, or including the assumption of the liabilities of the purchased company;
- shareholders of both companies are shareholders in each other in the same proportion after the event;
- the transaction is for "bona fide" commercial reasons and not for any form of tax evasion;
- then, any land transaction in relation to it, entered into as part of that transaction, is exempt from SDLT ("reconstruction relief").

25. This is a very limited relief and may be withdrawn if a disqualifying event occurs within three years (Schedule 7 of FA 2003 Para 7 (6)).

26. Relief is available where Company A acquires all or part of another Company, B, and the consideration is:

- only the issue of non-redeemable shares in a form similar to that set out in para 16 above
- or partly of such an issue, plus not more than the cash equivalent to 10% of the shares
- and/or the assumption or discharge of liabilities of the acquired company; (Schedule 7 to FA 2003, para 8)

27. Provided that there is no associated company of Company B involved with the shareholding in Company A, then the rate of tax payable on any associated land transaction as part of the transfer is fixed at 0.5%, as is normal for shares under the stamp duty regime ("acquisition relief"). The Finance Act 2005 introduced a measure in para 9 to Schedule 7 of FA 2003 restricting acquisition relief under para 8 to cases where the

undertaking (or part of an undertaking) acquired by the acquiring company has as its main activity:

- the carrying on of a trade, and
- that trade does not consist wholly or mainly of dealing in land or interests in land.

This measure also applies where the effective date of the transaction is on or after 17 March 2005.

28. Companies are "associated" under the provisions of s 416 of ICTA 1988; "arrangements" are defined in Schedule 7 to FA 2003, para 8(5) as including "any scheme, agreement or understanding, whether or not legally enforceable". Also, the relief can be withdrawn if a disqualifying event occurs within three years.

29. Either reconstruction relief or acquisition relief can be withdrawn on a similar basis to that set out in para 9 above if:

- the acquiring company changes because of an arrangement made before or during the three-year period
- the acquiring company or associated company has, at that time, a chargeable interest in the property asset (Schedule 7 to FA 2003, para 9).

There are some situations where reconstruction and acquisition relief are not withdrawn despite a change in control of the acquiring company. These are:

- transactions in connection with divorce
- transactions in connection with a testamentary disposition
- share transfers that are exempt under group relief
- share transfers to a company exempt from stamp duty under s 77 of Finance Act 1986
- because a loan creditor becomes, or ceases to be in control of the company.

30. In these circumstances where relief is withdrawn, a similar situation occurs as that set out in para 11 et seq above.

31. The meaning of "control" is set out in s 416 of ICTA 1998, together with the meaning of associated companies.

32. Schedule 7 to FA 2003, para 10, sets out circumstances where the withdrawal of relief is not appropriate. These are, basically, where transactions are exempt due to divorce, judicial separation or similar circumstances; where the variation is due to a death; where an intergroup transfer, exempt from stamp duty under s 42 of FA 1930, occurs; where an exempt company under s 77 of FA 1988 acquires the shares; or where the transaction involves a loan creditor ceasing to have control because of the payment of the debt or similar. However, in certain circumstances, the three-year rule set out in para 9 above applies.

33. If, where relief is withdrawn, the tax remains outstanding, then similar procedures to those detailed in paras 14 and 15 above apply (Schedule 7 of FA 2003, para 12).

Index